Canal Days in America

Canal Days in America

THE HISTORY AND ROMANCE OF OLD TOWPATHS AND WATERWAYS

BY
HARRY SINCLAIR DRAGO

 Clarkson N. Potter, Inc./Publisher NEW YORK

DISTRIBUTED BY CROWN PUBLISHERS, INC.

Printed in the United States of America
Library of Congress Catalog Card Number: 72–187510
ISBN: 0-517-500876
Published simultaneously in Canada by General Publishing
Company Limited. Inquiries should be addressed to Clarkson N. Potter,
Inc., 419 Park Avenue South, New York, N.Y. 10016.
First Edition
Designed by Shari de Miskey

Contents

Canal Days in America

Introduction

AT THE END of the first half of the nineteenth century the internal waterways of the United States—canals and rivers—were regarded as an important resource. In no European country, including Russia, would such an invaluable asset have been permitted to perish through neglect. But in our infatuation with speed and the railroads, we hastily abandoned our canals and consigned steamboats to a secondary role.

Today, however, it is the railroads that are hurting. Faced with ever-rising costs and feeling the pinch of air and highway competition, they have discontinued thousands of miles of once profitable branch lines. Gone are the Century, Super Chief and the score or more of deluxe mainliners, once the pride of American railroading. Ironically, this comes at a time when the tonnage being forwarded on the New York Barge Canal and the Ohio and Mississippi rivers continues to rise.

Obviously a mistake was made somewhere.

Recorded history does not reveal where, when or by whom the world's first canals were built. However, we do know that as long ago as 5000 B.C. ancient Babylonia and Egypt had discovered that a small scow or barge placed in a ditch filled with water and drawn by slaves was a cheap and practical means of transporting merchandise from one location to another. Although such primitive waterways

1

must have been very small, some fifteen centuries before Christ Babylonia's network of canals is said to have been largely responsible for her power and glory.

In Egypt, along the fertile central valley of the Nile, water was led off from the river when it was in flood to irrigate lands that would receive no other moisture during the year. In 1380 B.C. the Egyptians accomplished the feat of connecting the Mediterranean with the Red Sea. This channel was dug to facilitate trade and proved so successful that it remained in operation until 1859, when de Lesseps, the French engineer, began cutting the world-famous Suez Canal, on which thousands of men toiled for ten years.

A number of short canals had been dug in Northern Italy and adjoining France prior to the return of Marco Polo from the Orient, where he had been entertained at the court of Kublai Khan, Emperor of China. Among the many wonders he spoke of was the way the Chinese raised and lowered boats from one level to another. This, of course, was the lock, which the Chinese were using on their Grand Canal, one of the world's longest canals, connecting the Yangtze-kiang with the Pei Ho. Through the centuries that have intervened since then, the hydraulic principle by which a lock functions has not changed, only the lock gate has been improved. In the 1850s the drop gate, which rises and falls vertically, came into use on important American canals. It was not only a time-saver but had other advantages over the old cumbersome, manually operated swing gate. Its efficiency was further increased when it was electrified at the end of the century. However, in Europe, especially in France and the Low Countries, the old-fashioned swing gate is still widely used.

While Holland and Belgium have no canal as famous as the French Languedoc, which connects the Bay of Biscay with the Mediterranean, their system of interlocking short canals is the most efficient and valuable in the world. This has not happened by chance, for the phlegmatic Dutch and the practical Belgians realized long ago that the prosperity of their internal commerce could be insured by digging canals to connect their numerous natural waterways, large and small.

The flat terrain, the soil free of rock outcroppings, made digging a channel easy and inexpensive. In fact, as Alvin F. Harlow, the well-known canal historian, says, "The Low Countries were natural breeding grounds for canals." They are prosperous today, even venerated. One can only wonder what might have been the fate of our American canals if their worth had been similarly appreciated.

The prospect of wandering along an old towpath overgrown with weeds, searching for some remaining sign of the canal that ran there many years ago, arouses a pleasant nostalgia in the minds of many people. But where can one find an old towpath that will turn back the pages of time for an hour or two of dreaming? Of course, the physical evidence by which most of the old artificial waterways can be located and identified has all but disappeared. But there are almost countless local and county historical associations, as well as other sources of information, such as the Canal Society of New York State, that will gladly direct and assist a canal enthusiast.

If in your rambling you are fortunate enough to make a discovery, however minor, that other eyes have missed, your reward will be a thrill of satisfaction that will very likely turn you into a confirmed canal buff. In my own experience, I still recall vividly an incident that occurred on a June morning back in 1936 on the long-abandoned Black River Canal. Poking around in the soggy, brush-choked

Lansing Kill on the Black River Canal, made famous by Walter Edmonds in his novel *Rome Haul*.

channel in what I like to think of as "Edmonds Country," north of Boonville, I discovered one wing of a decaying lock gate hanging askew by one rusted hinge. While only a trifling matter, it has kept the Black River Canal green in my memory.

If you would like to spend a few tranquil hours traveling aboard an ancient canal boat, you can do so at New Hope, Pennsylvania, at various times during the summer and autumn months, on what was once the Delaware Division of the Pennsylvania State Canal System. New Hope is a pleasant town, and excellent food is served in its leading restaurants. Its large colony of writers, playwrights and artists, mostly refugees from New York, gives the place a holiday air. On the street you will overhear so much canal talk that you can acknowledge your interest in the old waterways without fear of being ridiculed.

It is only a few hours from New Hope down to Georgetown (in Washington, D.C.), where you can board a comfortable barge and be hauled up the old Chesapeake and Ohio Canal for the better part of a day. It is a rewarding experience. Among your fellow passengers there may be a senator or two or a

Peaceful, unpolluted stretches such as this on the Chesapeake and Ohio Canal invite the hiker and nature lover. *Courtesy National Archives.*

cabinet officer. Even presidents have been known to make the round trip on the canal for a few hours of relaxation from the cares of government.

But if you want to experience the pleasure of a leisurely passage on a working canal, make the Rideau Canal, in southern Ontario, Canada, your objective, for it functions today very much as it did 140 years ago when British Colonel John By and the Royal Engineers built it. At Kingston, on Lake Ontario, the Rideau begins its 125-mile northerly journey to Ottawa, the Dominion capital, on the Ottawa River. Almost from the time it leaves Kingston behind, it courses through a beautifully wooded pastoral country of farms and tiny villages. It is a region of countless small freshwater lakes. The Rideau Canal strings them together as though they were beads on a rosary. The canal is toll-free and has no commerce other than the hundreds of pleasure craft ranging from canoes and outboards to smart cruisers that follow its carefully marked channel every summer.

A few days on the Rideau will be sufficient to make you aware of what we lost when we neglected our own canals and permitted them to disappear.

1

How and Where It Began

B Y THE TIME the approximately eight-mile-long sea-level Cape Cod Canal was opened to navigation in 1914, the canal era in America was over and largely forgotten. It therefore does not come within the scope of a narrative dealing with "old" canals, with canals constructed between 1790 and 1850, when 4,400 miles of hand-dug artificial waterways—more than the distance from New York to San Francisco—were built. But Cape Cod must be considered, for it was there that the cutting of a canal was publicly debated by Americans for the first time.

History awards the distinction of having been the original advocate of a canal on this continent to Louis Joliet, the deservedly famous French explorer and fur trader. Joliet, son of a wagon maker, was born in Quebec in 1646 and was educated by the Jesuits, then at the height of their power in French Canada. At an early age he resolved to become a priest. He received the tonsure and minor orders. Four years later he renounced the clerical vocation and became a fur trader in the employ of the Hundred Associates, who then ruled the commercial life of Canada.

Joliet remained a half-priest for the remainder of his life. He rose into high favor with Jean Talon, the intendant (king's governor), who in 1673 appointed him to lead an expedition to discover and explore a possible water route to the great and largely unknown Mississippi River. Named to accompany him on his perilous journey was Father Jacques Marquette and five seasoned *coureurs de bois.*

Joliet joined Father Marquette at Pointe St. Ignace on the Upper Lakes, where the latter was in charge of the mission at that most important station. Of Marquette the great Francis Parkman has said: "He was a true counterpart of those earlier martyred missionaries, Isaac Jogues, Jean de Brébeuf and the gentle Garnier."

The party paddled through the Strait of Michilimackinac and, after skirting the northern shore of Lake Michigan, was off on its long wilderness journey. That they were ultimately successful in discovering the Mississippi, which they descended to present-day Memphis, is too well known to need comment. It is their return journey, when after weeks of fighting the adverse currents they reached the placid Illinois River, that is pertinent.

By easy stages they paddled up the Illinois for some 250 miles. On either side as they advanced, the level, boundless prairies stretched away to the horizon. Lake Michigan was their goal. Where today's statue of Joliet stands they encountered the Des Plaines River, a small feeder of the Illinois, which they followed for some fifty miles. When it began turning away from their objective, they left it. Heading north, a short portage brought them to a sluggish little stream, called by the Indians the Chicago River, down which they glided to the blue expanse of Lake Michigan.

It was while they were bivouacked on the Chicago River that Joliet pointed out to Marquette how easily water passage between Lake Michigan and the Illinois River might be achieved by digging a short canal. Evidence of this is found in the official Jesuit report of the expedition written by Father Pierre Dablon in 1674. It says:

> According to the researches and explorations of Joliet, we can easily go to Florida in boats and by a very good navigation with slight improvement. There will be but one canal to make—and that by cutting only one-half league of prairie from the lake of the Illinois [Michigan] into the St. Louis [Illinois] River, which empties into the Mississippi.

It can hardly be attributed to mere coincidence that when the state of Illinois built the Illinois and Michigan Canal in 1827, a century and a half later, it followed the course first envisioned by Joliet.

There is, however, definite evidence that cutting a canal across the neck of Cape Cod had been under active discussion for some time, possibly for as long as two years, when Father Dablon's report on the Joliet-Marquette expedition became available. The first mention is found in the diary of Samuel Sewell, a merchant of the village of Sandwich. On a cold, blustery October afternoon, with the whitecaps rolling in from across the bay, he and a friend rode along the sandy beach. Under the date of October 26, 1676, he wrote:

"Mr. Smith of Sandwich, rode with me and showed me the place where some had thought to cut, for to make a passage from the south sea [Buzzard's Bay] to the north [Cape Cod Bay]."

Whether the first man to contemplate the building of a canal in this country was Louis Joliet or some unknown New Englander—very likely a seafaring man if the latter—is of no great consequence. The fact that an enthusiasm for the canal continued to grow in Massachusetts is important, and although none was built for a century and more, efforts to pursue the matter were made periodically. In 1697 and again in 1776 the General Court of Massachusetts named committees to "view" the ground across the neck of the Cape and report their findings on the most likely location for such an undertaking. The gentlemen appointed "walked off" several routes. No surveys were made and no official estimates rendered on the cost of the projected undertaking. Private estimates were made, however, and they were so discouragingly high that public interest waned and could not be revived for years.

While digging an artificial waterway to connect Cape Cod Bay with Buzzard's Bay would have almost halved the distance by sea to New Amsterdam, the colony's most important market, and obviated the dangers of the outside passage around the tip of the Cape in foul weather, even the canal's most enthusiastic supporters realized that the cost of such an undertaking placed it beyond reach, for the population of what we know today as the Greater Boston area was less than ten thousand.

But by 1790 no fewer than thirty canal companies had been incorporated in eight of the original thirteen states. Some canals were already under construction, others were still in the planning stage. Caught up in a surge of prosperity, the country hailed with enthusiasm anything that came under the catch phrase of Internal Improvements. Before the period ended, the government was to invest $363,000,000 in canal construction. High as that figure is, 74.3 percent of the total investment in canals was raised by public subscription.

Most of the early canals were very short, seldom more than two or three miles long. They were not confined to one section of the country; between 1792 and 1796 canals were either being built or were in operation from New England down through states along the Atlantic seaboard to the Carolinas and even to distant Louisiana, the latter at that time a Spanish colony. To review them in chronological order would be confusing, for the dates of construction frequently overlap one another; dividing them geographically will better serve the purpose.

Public interest in a man-made waterway was first displayed in New England. Although the Cape Cod Canal had been moribund for years, it undoubtedly entered into the thinking of the Middlesex Canal Company and Judge James Sullivan of the Superior Court of Massachusetts, later governor of the state, its chief proponent. When the company petitioned the state for a charter in 1793, Governor John Hancock granted the request, giving to the "Proprietors of the Middlesex Canal Company the authority to construct a canal from Boston Harbor to Chelmsford on the Merrimack River, collect tolls and sell stock in the company."

Loammi Baldwin of the village of Woburn was elected to the office of superintendent.[1] The directors of the company could not have named a better man. Largely self-educated, he was fired with a mighty ambition and supreme faith in his ability to prosecute the undertaking to a successful conclusion. But neither he nor his fellow directors had anything but the vaguest idea of the course the canal should take in reaching Billerica, some twenty miles or more to the north, where it would cross the Concord River, which was to be its principal feeder, and continue on to Chelmsford on the Merrimack.[2] At that early stage little thought appears to have been given to selecting a southern terminus for the canal other than Boston harbor. In due course, however, the Middlesex reached salt water at Charlestown's millpond.

For all his admirable qualities as a leader, Baldwin was not an engineer and knew nothing about building a canal. At Harvard University library he had read the few books available on English canals, which had been in operation for the past thirty years, and he understood the principle by which a set of canal locks performed the miracle of raising or lowering barges at will, but he had never seen any. It is doubtful if he had ever seen a canal. Realizing that his ignorance and lack of experience placed him at a disadvantage, he began to cast about for a qualified engineer.

He searched in vain, for at that time there were not more than one or two experienced hydraulic engineers in the United States. Only one, William Weston, an Englishman who had lately come to this country, was widely known. His services were in such demand that he was currently dividing his time among four canal projects.

In his frustration Baldwin turned to Samuel Thompson, a magistrate in his home village of Woburn and a self-taught surveyor. Thompson, an earnest if ignorant man, met a committee consisting of Colonel Baldwin (a title won in the War for Independence), Judge Winthrop and several others at Medford Bridge late in October, 1793, and began his survey of the proposed canal, his only instrument a hand compass.

"A chaise accompanied us, carrying victuals and baggage," Loammi Baldwin noted in his diary. He continues:

> Up Mystic River to Mystic Pond, through the pond and up Symmes River to Horn Pond in Woburn, the survey progressed. At the head of Horn Pond apparently insurmountable obstacles stood in the way. Back we went to Richardson's mill on Symmes River and followed the valley through the eastern part of Woburn to Wilmington, finding an easy regular ascent to Concord River. The distance travelled thus far was about eighteen miles and the ascent, reported by the surveyor, sixty-eight and one-half feet. A year later, with an accurate survey, apparently this same ascent from Medford Bridge to Billerica Pond was found to be one hundred feet.[3]

This was not the only error. Thompson continued from Billerica on the Concord River to Chelmsford on the Merrimack and wrote in his report: "The water we estimate in the Merrimack at sixteen and one-half feet above that at Billerica Bridge, and the distance six miles." In truth, however, the water at Billerica Bridge was about twenty-five feet higher than the Merrimack at Chelmsford. Thompson's error in determining the elevation "was no less than forty-one and one-half feet in a short six miles." [4]

Thompson's miscalculations were too apparent to escape the eye of Loammi Baldwin and Judge Sullivan. They convinced their fellow directors that before work began, every effort should be made to induce Weston to come to Massachusetts and take charge of building the Middlesex Canal, no matter how extravagant his terms might be. It occasioned a great deal of traveling back and forth between Boston and Philadelphia, no light undertaking in those days, before the concerned parties reached an agreement and the Englishman began his labors for the Middlesex.

Construction began at Medford Bridge on September 10, 1794, Colonel Baldwin turning the first spadeful of earth. Prospects for the financial success of the Middlesex appeared bright as the Boston area received the major part of its timber, firewood, stone, cut granite and farm produce from New Hampshire's seemingly inexhaustible resources. Practically all of it was being floated down the

Merrimack River, but at Chelmsford the river changed its southerly course, turned eastward and reached its mouth at Newburyport, forty miles north of Boston by land and twice as far by sea. The expense of transferring merchandise from barge or raft to ships or wagons for the rest of the journey to the city increased their cost on the Boston market.

It was reasoned that the shorter, more direct Middlesex Canal would overcome that disadvantage, and even with the payment of tolls would be cheaper. Up to a point, this was sound logic, but it ignored the fact that New Hampshire was thinly settled and that for years to come the canal would be doing a largely one-way business. But the high caliber of the men behind the promotion convinced the public that it would be successful. When its stock was first offered for sale in 1794, it was pegged at $25 a share; in 1803 when full navigation had been established, the price rose to $475.

On the last day of December, 1803, Merrimack River water came through the Middlesex for the first time. At its highest point, Billerica Pond, the canal was 107 feet above tidewater. It crossed the Concord River at grade at Chelmsford, the mules crossing on a floating towpath. This was but one of William Weston's successful innovations. Five and a half miles north of the crossing it reached the Merrimack. Its overall length, not including the subsequent opening to navigation of the Merrimack to the village of that name, was 27½ miles.

Weston had built twenty locks, most of them stone, forty-eight bridges and eight aqueducts. The locks were twelve feet wide and long enough to accommodate boats measuring up to seventy-five feet in length and drawing three feet of water At Pawtucket Falls on the Merrimack, he found time to dig a diversionary canal, a mile and a half long, around what the Middlesex directors viewed as a promising millsite of the future. It chanced to be where the town of Lowell was to take root and prosper.

While Weston was spending money on improvements at a rate that made the thrifty New Englanders wince, Loammi Baldwin had five hundred men digging the canal. They were mostly local men toiling for $8 a month, the standard wage at that time for common labor, in addition to which they were housed and fed. Skilled masons and carpenters received $15 a month. Providing food and shelter for a work force of that size and supplying fodder for several hundred mules and horses, as well as constantly battling the weather, must have sorely tried indomitable Loammi Baldwin. A respite came when winter set in: the ground froze and work stopped.

As was the case with all the early canals, the Middlesex was hand-dug with axe, pick and shovel. For blasting, they had no more powerful explosive than black powder; Dupont's Improved Blasting Powder was not put on the market for another ten years.

English canal builders had demonstrated that a watertight stone canal lock could be constructed only with a generous use of hydraulic cement that hardened under water. It was composed of granite dust and trass, a pumice-like substance of volcanic origin. Great deposits of it were found along the Rhine in Germany, but none had been discovered in England or the United States. The Middlesex Canal Company found it very expensive to import.

A great to-do has always been made over the locating in 1818 of a great supply of trass at Chittenango, New York, which could be loaded on barges and floated to where it was needed by Canvass White, a young engineer engaged in building the Erie Canal. White's discovery is estimated to have saved the builders of the Erie from one-half million to a million dollars. It has often been claimed, though quite erroneously, that this was the first discovery of the substance in this country. It is a matter of record, however, that William Weston located a substantial deposit of trass within easy hauling distance of the Middlesex as early as 1797.

Weston left the Middlesex Canal Company soon after the canal was completed. Baldwin stepped down a year later and John Langdon Sullivan, son of Judge Sullivan, the original promoter of the Middlesex, took charge.

John Sullivan was an aggressive, determined man with an unquenchable optimism about the future of the Middlesex Canal Company, though it was then wallowing in debt. He foresaw a day when scores of textile mills would be drawn to the Merrimack with its abundant water power and furnish business for the Middlesex. Under his leadership a number of connecting small canals, commonly referred to as the Merrimack River Canals—the Wicasee, Cromwell's Pond, Union Locks and Canal, Amoskeag, Hookset and Bow—were joined and canal navigation to Concord, New Hampshire, established.

Not content to sit back and wait for the canal to attract business, Sullivan created business for it by building sawmills and gristmills and establishing a line of company boats to transport the produce to market. A good showman, he arranged junkets and feasts for prospective stockholders, taking them up the canal on the barge *Washington*, appropriately decorated with bunting and flags. Not stopping there, he inaugurated passenger service between Billerica and Chelmsford in gaily painted boats, the enthralled travelers seated on the deck beneath a striped awning. The fare was twenty-five cents for the two-hour journey.

Practically all of of the boats using the Middlesex Canal were flats, with a maximum capacity of twenty to twenty-five tons. There are old prints in existence showing them going down the canal, sometimes with a sail hoisted to aid the plodding mules. The Durham, the canal boat that came into being with the opening of the western canals, was not seen on the Middlesex until a few years before it passed out of existence.

Although the Middlesex was one of the earliest American canals it was one of the last to succumb to the competition of the railroads. While throughout its long life—fifty-seven years—its prospects were usually bright, it was seldom prosperous, largely due to the fact that the region it served did not make the industrial progress John Sullivan and others had expected.[5]

The Middlesex Canal cost $528,000 to build, a staggering sum for those years. Sullivan appealed to the state for aid but received only permission to conduct a lottery. The money it brought was soon expended. Once again the Middlesex stockholders were called on for another assessment. They paid it, and another and another. When the final assessment was made in 1817, it was the hundredth they had been called on to meet. By then the total assessments against each share of stock amounted to $740.00 and in return had paid $559.50 in dividends, which does not cover the loss of interest on the money invested.[6]

The end came on April 4, 1860, when the property and assets of the Middlesex Canal Company were ordered seized and forfeited. At the forced sale only $130,000 was realized, the Boston and Lowell Railroad being the principal purchaser. The money was divided among the stockholders, this final dividend permitting them to come out even.

2

The Connecticut River Canals

TODAY MOST AMERICANS, including the young, know that what is called the *Old* Boston Post Road was once the principal thoroughfare between Boston and New York City. Few are aware, however, that there were three Boston Post Roads—branches, if you like. There was the Upper Road, which reached the Connecticut River at Springfield from Boston and Worcester. The Middle Road ran through Dedham, Uxbridge and Pomfret to Hartford. The third, or Lower Road, struck off for Providence on leaving Dedham and followed the Connecticut shore to New Haven, where the three roads became one for the rest of the way to New York City.

The Upper Road, after reaching Springfield, turned southward along the Connecticut River to Hartford, where it merged with the Middle Road and they continued as one to New Haven. Hartford dominated the rich Connecticut River Valley and was that river's busiest port.

Today, we may not think of Hartford as a thriving river town, with competing steamship lines giving it daily freight and passenger service to New York City. Nor in the summertime, when the harbors from Haddam down to Essex are crowded with slick yachts and the old Essex House comes alive again for a few days and nights, is it easy to recall the days when river traffic was restricted almost exclusively to freight and passenger vessels.[1]

The Connecticut was not an easy river to navigate. In the twenty-one miles between Middletown and Hartford to the north, there were numerous bars and other impediments. The channel was narrow, with an average depth of no more than 5½ feet. Following a freshet, it often shifted from one side of the river to the other. In 1800 the Union Company was granted a sixty-year charter "to improve Navigation of the Connecticut River," for which it was to be permitted to charge tolls on all boats using the river between Middletown and Hartford. "It dredged the bars, removed obstructions, erected piers, stoned banks and planted willows to protect the banks against rushing flood waters." [2]

This was one of the happiest arrangements ever made between a private corporation and government. The company continued to operate for the duration of its charter, in the course of which it expended $34,000 in deepening the channel and on other improvements, an additional $12,000 being contributed by the city of Hartford.

Hartford was regarded as the head of navigation on the Connecticut, but small stern-wheelers never stopped trying to navigate the upper river, being poled through the rapids by a score of so-called swift-water men or dragged by teams of six to eight oxen operating at the end of a hundred-foot chain.

One of the last of the little stern-wheelers to get above Montague Falls was *Brownstone*, captained by Edward King, who, in the parlance of the times, was "quite a character." The *Brownstone* went aground in Long Meadows. Captain King jumped overboard and found himself standing in two feet of water.

"Hellsfire," he shouted to the mate, "she ought not to be grounded here; it's knee deep. You better put out a lantern."

"What for?" the latter asked.

"What for?" the exasperated captain shot back. "Why some farmer might try to drive across here and there'd be hell to pay."

The tale occurs often enough in the folklore of the river to give it some validity. In its rustic humor it reflects the difficulties of navigating the upper Connecticut. Obviously it was to overcome those difficulties and provide comparatively free passage that a series of short, homemade, Yankee canals was built, most of them antedating the steamboat.

The first—and it was New England's oldest and second shortest man-made waterway—was the South Hadley Falls Canal, a few miles downriver from the present city of Northampton, Massachusetts. Work on it began in 1793 and was completed the following year. Although no more than two miles long, it was important to the navigation of the upper Connecticut.

It was built by men who from long experience with poling flats up that stream and portaging around the falls were well acquainted with its tantrums. But they were not engineers; and having no technical knowledge to assist them in devising a means of lifting a canal boat from the lower level of the river over the falls, a matter of fifty feet, they had to depend on their native ingenuity, which proved to be more than equal to the task.

In fact, upon completion, the South Hadley Falls Canal was regarded as a noteworthy achievement. English visitors, familiar with the canals of their own country, where the topography of the land made deep cutting unnecessary, regarded with wonder the vertical cleft forty feet deep and three hundred feet long that had been cut through solid rock.

In 1926, Alvin F. Harlow wrote:

> The little South Hadley ditch was noteworthy also because it built the first inclined plane in America—two hundred and thirty feet long with a vertical lift of fifty-three feet. The face of the plane was stone, covered with heavy plank. The body of the car (which was raised and lowered at will) was a water-tight box with folding gates at each end. Two water wheels sixteen feet in diameter on either side of the channel at the head of the plane were operated by water from the canal, and pulled the car up or let it down, according as the gears were shifted. Boats floated directly into the car; the gates were then closed behind it and the car emptied water through sluices at the sides. The carriage was then pulled up or let down the plane on three sets of wheels, like big wagon wheels, graduated in size so as to hold the car exactly level.[3]

The inclined planes used many years later on the important Morris Canal in New Jersey operated on the same principle although deriving their power from a different source.

The South Hadley Falls Canal was at once profitable, but the canal company soon became the target of a campaign waged by fishermen who claimed that its dam prevented the shad and salmon from making their annual run up the river to their spawning grounds. From another direction came the outcry of farmers up the river that the dam was flooding their lowlands, causing serious outbreaks of malaria. The town of Northampton joined the ranks of those opposed to the South Hadley Falls Canal Company and in 1800 petitioned the legislature for the removal of the dam. The company countered by offering to lower the dam and eliminate the inclined plane, in return for the privilege of conducting a lottery in the amount of $20,000. This was done.

The remodeled dam was destroyed by a spring flood two years later. A second was destroyed in like manner, and in 1823, a third dam was swept away. Coupled with this latest disaster was the distressing fact that upriver business was declining. Great rafts of timber were coming downstream, but an increasing share of the logs were being cut into lumber at the mills at Mount Tom. To the south, Springfield was growing so rapidly that most of the upriver business ended there. As a consequence the canal company decided not to go to the expense of building still another dam, and restricted its operations to supplying water power to the small mills along its raceway and channel.

At Bellows Falls, Vermont, a canal only a mile long was cut around one of the meanest stretches of the Connecticut River. Short as it was, its nine locks overcame a drop of fifty feet and opened navigation upriver for 120 miles. The operation was so successful that enthusiasm was aroused for a through water route all the way to the St. Lawrence, the proponents of the idea claiming that on leaving the Connecticut at White River Junction, with minor improvements boats could make use of the third branch of White River to within a few miles of the Winooski at Montpelier, and by using Winooski have passage all the way to Lake Champlain. The ascent from the Connecticut to Montpelier would have necessitated installing at least fifty locks. Fortunately for the would-be investors this costly dream died a-borning.

Windsor Locks on the Connecticut River near Hartford, Connecticut.

The Montague Falls Canal and what was known as "the Turner's Falls reach" were built in 1800. Although only three miles long, it required eight locks and two dams to make the climb around one of the wickedest stretches of water on the Connecticut River. Although the canal was costly to build, the company seldom failed to declare an annual dividend, the bulk of its tolls being collected from the endless stream of timber being rafted downriver from upper Vermont and New Hampshire.

The last to be built and the longest and most important of the Connecticut River canals was the six-mile-long Windsor Locks Canal, twelve miles upriver from Hartford. It was opened to navigation in the middle 1820s, its purpose being to provide a safe passage around the wicked Enfield Falls and rapids. It has survived all other New England canals and is used today by countless pleasure craft, as well as to supply the factories at Windsor Mills with water. The old towpath remains and canal buffs will find it little changed by the century and a half that has passed since it was first laid.

Prior to the building of the canal, Warehouse Point, at Windsor, was virtually the head of navigation on the Connecticut River. It was fondly believed by many men that with the completion of the canal Windsor would be transformed into a booming inland seaport. While that fanciful dream was never realized, the canal did change the life of that peaceful section of the Connecticut River Valley as nothing else could have done.

The late Stewart Holbrook, the most knowledgeable of our grass-roots historians, observed:

> What the Locks did was to bring all river traffic to a halt, briefly, and procide taverns on both sides of the river to cater to the captains and crews of the boats, to the passengers, and drovers, and the army of river drivers accompanying the rafts and booms of logs that were cut far up the river in Vermont and New Hampshire.[4]

If you have ever seen a crew of reckless men, catlike in their movements, running and leaping on the rolling deck of a million feet of spruce as they rode it downriver, you can understand their exuberance and violence when it came time to play or fight. They worked hard, drank hard and fought the same way. At Windsor Locks they met their match in the equally reckless powdermen whose job it was to take downriver untold tons of explosives from the Hazard Powder Works at Hazardville, several miles from Enfield. The powder was packed in twenty-five-pound kegs and twelve hundred kegs was the load usually stacked on a barge. Few accidents occurred but the possibility of one was always present. Working that close to instant death toughened a man considerably and frightened away all but the hardiest.

Rival settlements sprang into existence at Windsor Locks, Windsor Hill and Point Rocks, with low groggeries, rowdy dance halls and dens of harlotry doing business cheek by jowl. A widely circulated saying had it that there was little a man wanted that could not be found at Windsor Locks. Overhead floated the intoxicating fumes given off by the distilleries at Warehouse Point.

At Warehouse Point you are in the heart of the tobacco region where for a century and a half the finest broadleaf produced in this country has been grown. According to legend, cigar-making in America began in nearby Suffield. The story has it that General Israel Putnam, Connecticut's favorite folk hero, returned from a trip with "three donkey-loads of Havana cigars" that met with such a warm reception that Simeon Viets, a tobacco grower in nearby Suffield, imported a professional from Havana to show the women of Suffield how to roll a cigar. Up to that time cigar-smoking had not been thought quite respectable. But Viets' cigars caught on immediately with tavern frequenters. His deluxe brand was known as Windsor Particulars. They were too expensive for most smokers, but his Short Sixes or "Twofers" were within reach of everyone—costing not two for a nickel, as you might suppose, but two for a penny.

3

The Blackstone
and
Other Yankee Canals

IN MAINE, four great rivers, the Androscoggin, Kennebec, Penobscot and St. Croix, flowed southward to the sea out of the vast and largely unknown timberlands to the north. General Benedict Arnold and his troops had fought their way through that wilderness of spruce and fir in their unsuccessful attempt to capture Quebec in 1775. The "Big Woods" they called it with bitterness. Half a century later it was still a wilderness, the millions of feet of timber that had been sent hurtling down the turbulent rivers to salt water having hardly scratched its abundance.

Maine was still sparsely populated. With its network of great rivers, the like of which no other state could match, all flowing in the right direction for avenues of commerce, it appeared to be the last state on the Atlantic seaboard to have need of a canal. And yet, under the title of the Cumberland and Oxford Canal Company, such an undertaking was organized and chartered in 1820.

The state of Maine granted the promoters a lottery privilege of $50,000. A few days later a group of Portland businessmen, most of whom were among the incorporators of the canal company, launched the Canal Bank, capitalized at $300,000, one-fourth of which was invested in the Cumberland and Oxford Canal Company.[1] Why this transparent bit of financial legerdemain was deemed necessary remains a mystery. Perhaps the intense rivalry between Portland and thriving

Bangor explains it. The latter was making such rapid strides that some Portlanders were expressing the fear that in another twenty-five years Bangor would be Maine's leading city.

The ostensible reason for building a canal from tidewater to Sebago Pond —"pond" being a localism for what would have been a lake elsewhere, and a large one—20½ miles to the north, was to open up a great slice of back country to commerce. Since Portlanders felt that the region about to be touched by the magic wand of progress was peculiarly their own, it may be assumed that they regarded the Cumberland and Oxford Canal not only as a sound investment but also as a slap at Bangor.

Work on the canal was begun in 1828 and was completed two years later. Subsequently, with the building of a lock in Songo River, navigation was extended to Brandy and Long ponds, giving the little Oxford an overall length of thirty miles. The timber and other forest products that it brought down to tidewater, and the lively business it conducted with the logging camps in the interior, made it prosperous from the start, and it remained so until the Maine Central and other railroads invaded its territory and put it out of business.

Another Yankee canal of an earlier period was the Blackstone, or Blackstone River Canal, connecting Narragansett Bay in Rhode Island with the city of Worcester, Massachusetts, forty-five miles to the north, which made it the next to longest canal ever dug in New England. In importance it ranked second to none and it enjoyed the experience, rare among canals, of doing a two-way business, its northbound tolls being matched, or nearly so, by the traffic in the opposite direction.

As early as 1796, only two years after work began on the Middlesex, a proposition was advanced by a group of Providence businessmen to connect Narragansett Bay with Worcester, which aroused statewide interest. A corporation was formed and quickly received a charter from the Rhode Island legislature. However, a similar petition to the Massachusetts legislature was promptly denied. This was said to have been due to the opposition of the rich Boston Brahmins who were heavy investors in the Middlesex Canal and were not disposed to countenance any possible competition by Providence and Worcester interests.[2]

The refusal of the Massachusetts legislature to grant a charter effectively killed the Blackstone project for years. But Rhode Islanders never lost interest in it, and when it was revived in 1822 and sanctioned by both states, the company's stock—$500,000—was subscribed several times over within a few hours after it was offered to the public.

Outstanding good fortune attended the building of the Blackstone Canal. The accidents, mistakes and ignorance from which similar undertakings suffered and which accounted for interminable delays did not occur. Excellent use was made of the Blackstone River, but the fact that forty-eight locks of cut granite had to be installed in a canal whose overall length was only forty-five miles is proof that the builders did not have an easier encounter than was the case on other canals of its size. Good management as well as good luck made the difference.

Ground was first broken in Rhode Island in 1824. Construction on the Worcester end did not begin until two years later. The first boats passed through the completed canal in September, 1828.

The Blackstone Canal did a prosperous business from its very beginning, and it continued to pay dividends for years, until the Providence and Worcester Railroad put it out of business in 1848. Its best year was 1832, when the tolls amounted to almost $19,000.[3]

It was on the Blackstone that canal boats began to be designed along lines that were to become classic among boat builders on the New York State and western canals. It was also on the Blackstone that canallers began to regard themselves as a segment of society apart from other men and draw together.

The Blackstone has the distinction of having been one of the first passenger-carrying canals in the United States. The calculated running time from end to end was twelve hours, which meant that you could leave Providence in the early morning and be in Worcester that evening. No accommodations were provided for passengers. They rode on deck and had to brave the elements and brought their lunch baskets with them. On approaching a bridge the captain raised his horn to his lips and sounded a warning. The ladies lowered their parasols and the gentlemen doffed their tall beaver hats.

Boatmen regarded children as a nuisance. With no deck railing to restrain them, they often fell overboard and had to be pulled out of the canal, unharmed but somewhat the worse for wear.

With the coming of winter, operations were suspended for three or four months. At other times, high or low water interfered with the slack-water navigation of the Blackstone. The canal company shared the water with the mills along its banks. When there wasn't enough water for both, trouble developed on more than one occasion that erupted into violence. Mill hands dropped rocks into the locks at night. The canallers responded by threatening to burn the mills. The situation became so tense once or twice that armed guards had to be posted to protect the mills.

The Blackstone never entertained any grandiose ideas of building into Boston harbor or anywhere else; it had its own little bailiwick and was content to keep within its borders. That was not the case with the Farmington Canal, a bit to the west in central Connecticut. As originally chartered by James Hillhouse in 1822, it was to have been a modest venture, connecting the little town of Farmington with New Haven, and at Farmington going up the little Farmington River as far as his money and good luck would take him. That was not the concept of the group of New Haven bankers, politicians and businessmen who took over the reins.

It was their announced intention to build a canal reaching from Long Island Sound, at New Haven, to Canada, connecting with the as yet unbuilt Hampshire and Hampden Canal at the Massachusetts line, going up the Connecticut through Vermont to Lake Memphremagog, then taking the St. Francis River northward to the St. Lawrence.

To dismiss this plan as absurd, fantastic propaganda is to flatter it. One can only wonder how dupes with money to invest were taken in by it. Of course, no attempt was ever made to build such a canal, but perhaps it affords a clue to the squandering of millions of dollars in other ways.

New Haven had Yale University, which gave it great cultural prestige. But it wanted more: it wanted its share of the business that its booming archrival Hartford had captured. The Great Farmington Canal, it was predicted, would be the catalyst that would secure it. The promoters, under the title of the Hampshire and Hampden Canal Company, secured a charter from Massachusetts in 1823, granting them the right to build a canal running a bit southwest from Northampton on the Connecticut River through Easthampton, Southampton, Westfield and Southwick to the Connecticut line. The Hampshire and Hampden Canal Company and the Farmington Canal Company were indistinguishable. The course the latter was to take led from Southwick southward to Granby, Simsbury, Farmington, Plainville, Southington, Hamden and New Haven. This route was designed to give Hartford a wide berth, and it skirted past the city at a safe distance, actually between seven and eight miles.

The Hartford *Courant* delighted its readers by taking jibes at New Haven and its "little ditch": "We have been thinking of applying for the bull frog concession on the canal, but the tailless amphibians require water. So we will wait until the spigot is turned on and we see what happens." The New Haven *Chronicle* returned fire with fire, and this editorial sniping whipped up support for the canal and put the town squarely behind the Farmington Canal Company. How deep the feeling went is reflected in the fact that the new Mechanics' Bank had to subscribe for $200,000 worth of canal company stock before it could be chartered. The city of New Haven and another bank obligated themselves for a like amount.

With appropriate ceremonies and speech-making, the digging of the Farmington Canal began at the state line on July 4, 1825, Governor Wolcott of Connecticut turning the first spadeful of earth. (In those years July 4, with its patriotic overtones, was deemed the most appropriate day of the year for getting such projects off to a good start.)

The old Farmington Canal after leaving Ten Mile River.

Breaking ground at the state line was only a symbolic gesture; actual digging began north of Farmington, to progress to the south. At the same time a work force began digging northward from New Haven. By the time they met and navigation was opened between Farmington and the city, it was obvious that completing the project to Northampton would cost many times the estimated figure, a miscalculation that confronted most canals.

The completed divisions of the Farmington soon developed a modest freighting business, carried machinery and general merchandise into the interior and brought out farm products, some lumber and hides. Its passenger business exceeded expectations. People living in the towns and farms along the canal evidently enjoyed traveling back and forth to the city by water. The boats were comfortable and, although not large, were the best yet to appear on New England canals. Sometimes they served as temporary hotels, as is indicated in the following item from the New Haven *Chronicle* of September 2, 1829; "The elegant Canal Packet *New England* will, for the purpose of accommodating passengers, leave Farmington on Tuesday before Commencement, which takes place on Wednesday, the 9th inst." [4] After attending the exercises (at Yale), it goes on to say, "Passengers will be permitted to spend the night on the boat, prior to an early departure for way points up the Canal."

The Farmington suffered more than the usual number of misfortunes. Twice the big stone arch that carried it over Salmon Brook at Granby was washed away. Spring freshets often destroyed stretches of its berm and towpath. Breaks in its banks due to flooding resulted in the payment of heavy damages to farmers whose lands had been inundated.

It took the Farmington ten years to reach Northampton, some seventy miles from New Haven, which made it the longest of all New England canals. When the first through boat from New Haven to Northampton, appropriately named for the latter city, journeyed up the canal carrying two hundred carefully selected passengers, gay with flags and buntings, it was greeted along the way by cheering crowds and thundering cannon. Horses were changed frequently and a speed of four miles an hour was maintained. When the last stretch of the journey was reached, four gray horses took the towline, and the *Northampton* reached the South Basin in style. From the bow of the *Northampton* Governor Wolcott waved to the governor of Massachusetts, who was on hand for this gala occasion. Cannon boomed and there were speeches. That night a banquet was tendered the governors and the builders of the canal, as it was believed the canal would pour gold into New Haven and Northampton.

In the morning, the *Northampton* was locked into the Connecticut River, and there were more speeches and hurrahing. But the gaiety and jubilation were a fraud, for the Farmington Canal Company was in the direst financial straits. To take in some quick money it reduced the fare for the twenty-four-hour run from New Haven to Northampton to $3.75 with meals. The boats were soon crowded with late summer travelers taking advantage of the bargain. But it was going to take bigger money than that to save the Farmington Canal Company.

The damage done by a flash flood in mid-September produced a crisis. The old company could not continue. A new corporation, the New Haven and Northampton Canal Company, took over the property, paid nothing for its

The Chesapeake and Delaware Canal in 1865. *From Clowes,* Shipways to the Sea.

The Chesapeake and Delaware Canal in modern times. *From Clowes,* Shipways to the Sea.

America for a canal. The line they staked ran from Middletown on the Susquehanna to Reading on the Schuylkill.

It revived talk of connecting the Susquehanna with the Schuylkill and in 1791 two companies were chartered for that purpose by the state of Pennsylvania, one to "build a canal from Middletown on the Susquehanna to connect with the Schuylkill River, and the other to open canal navigation on the Schuylkill from Reading to Norristown and build a canal from Norristown to the Delaware."[1]

In the following three years fifteen miles of canal were built, and a number of wooden locks constructed to lift it up the ascent from the Susquehanna. Four hundred thousand dollars had been spent. Funds were exhausted. The state came to the promoters' aid, granting them a lottery privilege of $400,000, the Schuylkill and Susquehanna Company to receive two-thirds of the proceeds and the Delaware and Susquehanna one-third. Although thousands of dollars worth of tickets, were sold, the canal companies received a bare $60,000. This discrepancy brought about the first disclosure of the extent of the lottery swindles. (A later chapter will be devoted to the lottery scandals.)

Although unconnected with the canal project, money was raised for the removal of the serious obstruction to navigation of the Susquehanna at Conewago Falls, just above the little town of Columbia, in 1834. The so-called falls were actually more a series of dangerous rapids. Local folklore invested them with tales of tragedy and disaster, which gained credence from the fact that hardy, experienced keelboat men coming downriver from the north stopped at Middletown, twenty miles upriver, and forwarded their furs and peltries to Philadelphia by wagon rather than risk running the rapids.

A mile-long canal was dug around the bad water at Conewago, and with the addition of sluices and the necessary sluice gates, a fifteen-mile stretch of treacherous white water was tamed. It was a solid achievement and it functioned until the Susquehanna and Tidewater Canal was built forty years later.

The success of that operation restored a measure of public confidence in the Schuylkill Canal and enabled it to limp along for years between long periods when no work at all was done. In 1811 a group of Philadelphia bankers and wealthy citizens informed the harassed directors that if the financial affairs of the two companies were placed in their hands, they would not only advance money for completion of the project but also guarantee that the lottery privilege would be renewed for another five years, provided the two companies merged and a competent engineer was put in charge of construction. This offer must have carried a price tag, but there is nothing in the record to disclose what it was.

Those terms were agreed to and the old companies merged into the Union Canal Company. Loammi Baldwin, Jr., the worthy son of the builder of the Middlesex Canal, was brought down from Massachusetts and installed as chief engineer.[2] Under his direction the work went forward at a quickened pace. But his insistence that dams and feeders would have to be built to provide the canal with the water it would need led to a series of disagreements that ended in his dismissal. In 1824 young Canvass White, who was making a name for himself as chief assistant engineer on the Erie, the yardstick by which all canals were to be measured, took charge.

In passing, it is worth noting that Baldwin's assessment of the Union Canal's

Weighlocks, c. 1880, on the Pennsylvania Canal. *Courtesy Edwin P. Alexander Collection.*

Coal boats on the Schuylkill Canal.

water needs proved to be correct. Lack of water was to be the curse of what came to be called the Schuylkill Navigation.

The Schuylkill Canal, or what may be termed the eastern division of the Union Canal, did not have to go looking for business. In 1826, a year before it was regarded as complete, it was opened to navigation from Philadelphia to Mount Carbon, just above Pottsville at what became Port Carbon, a distance of approximately 108 miles, that was accomplished in almost equal parts by slack-water navigation, created by numerous dams, and a hand-dug ditch. The gross tonnage carried that year is said to have totaled more than a thousand tons. At Pottsville the canal was 588 feet above sea level.

The first boat to be launched on the Schuylkill was built either at Orwigsburg or Schaefferstown and hauled to the water by oxen.[3] Architecturally it set the design for the superstructure of the canal boats that were to ply the Erie and all other later canals. It had a rounded bow, a blunt stern, and was steered by a sweep or rudder. There was a low cabin in the rear, rising not more than thirty inches above deck level. Undoubtedly the handsome boats that appeared on the Blackstone Canal were inspired by it.

In preparing to build a canal, it was seldom the primary rule to establish the shortest possible distance between two points and follow it. Instead, you took the cheapest way of getting to where you were going, avoiding as many costly fills and cuts as you could. The added miles were of no great consequence, for the pace of life was leisurely. No better illustration of this can be found than in the wandering of the Susquehanna or western division of the Union Canal before reaching the eastern branch at Pottsville. By way of Lebanon and Pine Grove it pointed northeastward from its starting point at Middletown on the Susquehanna. When it arrived at Pottsville it was almost as far from Reading as it had been at Middletown.

To the Schuylkill Canal builders must go credit for boring the first American tunnel. It was dug through Mount Carbon, which was a ridge rather than a mountain, and was four hundred feet in length. When completed in 1821 it was heralded as one of the engineering wonders of the day. People came long distances to see it and be thrilled by riding underground through a "mountain." Aside from the publicity it attracted, there appears to have been little reason for building it. Mount Carbon could have been avoided by a slight deviation in the route. The tunnel was shortened a few years later and made into an open cutting in 1857.

A much larger tunnel, 729 feet long, 18 feet wide and 16 feet high, was bored on the Susquehanna division. Boats entered it through handsome portals of cut stone. Although costly to build, the expense was justified, for it took the canal over the rocky summit of the watershed that otherwise could have been surmounted only by a great number of locks. It was shortened somewhat a few years later when the old works of the original Schuylkill and Susquehanna were largely abandoned.

In 1827, sixty-five years after the original survey, the Union Canal was declared open to navigation from end to end. The *Fair Trader*, the first boat to reach Middletown from Philadelphia, made the passage in five days, its progress limited to the daylight hours. It was an achievement that silenced the doubting Thomases who had been saying for years that the canal was only a scheme to extract money from fools and would never work. Having been alerted in advance to the coming of the *Fair Trader*, Middletown greeted its arrival with what was described as "a monster celebration." [4]

If there is a time for celebration, there is also a time for despair, and the Susquehanna division of the Union was faced with the latter when it was discovered that its locks, only 8½ feet wide and 75 feet long, were too small to accommodate the large boats that were appearing on the Schuylkill, where the channel was twice as wide and the locks 90 feet long.

That this was not discovered long before the canal was in operation seems incredible. To overcome the dilemma the building of small boats was encouraged. The campaign was so successful that by the end of the year an estimated two hundred small boats had been launched. But they were not capable of carrying more than twenty tons of freight. This meant that every boat locked through represented an average loss of eight tons in tolls.

To overcome this deficiency the locks were kept in almost constant operation, which resulted in a tremendous loss of water. Baldwin's contention that the canal would have to develop an auxiliary water supply could no longer be doubted. The problem was solved by building the great forty-five-foot-high Swatara Creek dam and a giant pumping station to put the water where it was needed. In addition, a navigable twenty-two-mile-long feeder was dug to Pine Grove. These improvements were completed by 1830 and widening of the channel and locks was begun.

Under the terms of its charter the Union Canal Company was obligated to connect the Schuylkill with the Delaware River. Doubting that the seventeen-mile connection between Reading and Bordentown would be profitable, it put off construction until it had exhausted its legal excuses for further delay. Unwittingly, this link, forged so reluctantly, was to bring the Union its greatest prosperity.

From German kitchens certain gustatory delights usually not available went down the canal to Philadelphia. Back came, among other things, catalogues and newspapers showing the latest female fashions and underthings being displayed in Broad Street shops. Mennonite women shuddered at such boldness and wondered what the world was coming to. Behind their mothers' backs, Mennonite girls giggled and were envious.

By 1840 the Union Canal represented an investment of $6,000,000. It had become one of the busiest canals in the country. But obviously nothing less than a miracle was going to be required if its long-suffering stockholders were to be rewarded with a profit.

Although not yet appreciated, a miracle was already occurring, for in addition to the farm products and timber that were the mainstay of most canals, the Union Canal was forwarding a skyrocketing tonnage of once despised and worthless stone coal—Pennsylvania anthracite—a miracle that was to make millionaires of paupers.

If Philadelphia and Philadelphia money exerted a tremendous influence on the advancement of the country in the years preceding the War for Independence and for several decades thereafter, it is not difficult to understand why, for it was the largest and most commercially important city in America. It was interested not only in expanding its trade to the West but in improving it in the opposite direction. With understandable enthusiasm it encouraged talk of cutting a canal across the narrowest section of the three-hundred-mile-long peninsula that separated Delaware Bay from Chesapeake Bay. Such a facility would shorten the week-long voyage around Cape Charles to Baltimore, Philadelphia's principal rival, to a matter of hours.

In 1764, only two years after Rittenhouse and Smith made their historic survey of what was to become the route of the Union Canal, a preliminary survey —really only a walk-off without instruments—for a waterway connecting the Delaware and the Chesapeake was made. Nothing came of it. In 1769, however, the Philosophical Society of Philadelphia raised £200 (somewhat less than $1,000) from the businessmen of that city to defray the cost of a reliable survey to connect Chesapeake and Delaware bays.

To take advantage of what navigable water there was on the little Apoquimene River, which flowed into the Chesapeake, and the Bohemia River, discharging eastward into Delaware Bay, leaving only the rocky spine of the verdant peninsula in between to be crossed, the survey took that course. When the report was made to the Philadelphians it was rejected, the contention being voiced that a canal that far north would be of greater advantage to rival Baltimore than to the Quaker City.

Interest in the project dwindled and with the coming of the Revolution it disappeared completely. But it was a sound idea and its benefits too apparent for it to remain dormant forever. When the strains imposed by the long war were past, interest in a Chesapeake and Delaware Canal revived. But it was not

The Chesapeake and Delaware Canal, a continuing operation for the Corps of Army Engineers. *Courtesy National Archives.*

St. Georges, original lock on Chesapeake and Delaware Canal.

until 1802, with Maryland joining Pennsylvania and Delaware in chartering the Chesapeake and Delaware Canal Company, that work began.

The public subscribed for $400,000 of its stock, of which perhaps as much as 40 percent was paid in. Hopes ran high that the canal could be completed in two or three years. It was to be a short canal, approximately twenty miles from bay to bay. By making use of the navigable water of the two little rivers, the miles of canal to be dug would be much less than that. However, at the end of the year, when it was revealed that $100,000 had been wasted on needless surveys and other preliminaries, the bottom fell out of the promotion and stockholders refused to pay the assessments. The company failed and the project was abandoned.

For the following ten years the Chesapeake and Delaware lay moribund, seemingly without hope of ever being revived. Then in 1822 the state of Pennsylvania pumped life into it. Admittedly it was acting in its own self-interest, but it was the only source from which the push to restore confidence in building the canal could have come. Financially, Delaware was in no position to do it nor was Maryland (meaning Baltimore) ready to open the Chesapeake Bay trade to Philadelphia.

By presenting the project as an internal improvement affecting the prosperity of the three states, Pennsylvania, the richest and politically most powerful of the former colonies, succeeded in winning the endorsement of the federal government and a committment of $500,000. Pennsylvania contributed $100,000 and induced Maryland to underwrite an obligation of $50,000. Little Delaware followed with a modest $25,000. With confidence restored and interest aroused as never before, private investors swelled the total money committed to over one million dollars.

The assets and title of the old company were purchased. The first task of the new board of directors was to appoint a permanent engineer. Their choice was Dr. Benjamin Wright, then in charge of the middle section of the Erie—Seneca River to Rome. Wright refused to leave his post. The second choice of the directors was John Randle, Jr.

They could not have done better. Randle had been previously employed as an assistant engineer by the original Chesapeake and Delaware Company and been dismissed for insisting that the route being taken would not provide enough water to get the canal over the backbone of the peninsula, while a course farther south would.

Randle had it his way this second time around. He shortened the length of the canal to 13⅝ miles, but at a terrific price in men, money and time, his grit and determination being no substitute for the engineering skill he did not possess. He had not contemplated building dams to supply the canal with water. Three were found to be necessary. Nor did he have any conception of the difficulties he would have in cutting through the backbone of the peninsula, which proved to be a barrier of solid granite. Driving a passage through it for a mile and more, at depths ranging from seventy-six to ninety feet and sixty feet wide, was to be the work of years. At times, as many as twenty-five hundred men labored, using primitive tools. Black powder was still the only available explosive.

This was the famous Deep Cut, regarded as one of the engineering triumphs of the time. Immense rock slides maimed or fatally injured a score of workmen.[5]

The work force was recruited in Philadelphia and was mostly composed of

The main channel of the Chesapeake and Delaware Canal westward from St. George's Bridge. *Courtesy National Archives.*

The Deep Cut on the Chesapeake and Delaware Canal. *Courtesy Inland Waterways.*

Old drawbridge on the Chesapeake and Delaware Canal. *Courtesy Inland Waterways.*

Corkonians lately arrived from Ireland. On paydays they dropped their tools and hurried off to the shantytown on the bayshore that had been set up to cater to their appetite for whisky and women.

The expense of clearing away thousands of cubic yards of rock and earth helped to send the cost of the enterprise soaring. Long before it was finished it was apparent that the Chesapeake and Delaware Canal was going to be the costliest canal yet dug in the United States.

Seven miles of the unfinished eastern section were opened to navigation in 1828 and used by sailing sloops and such shallow-draft steam-packets as the *Lady Clinton* of Wilmington. On October 17, 1829, water was turned into the Deep Cut.

The first year the Chesapeake and Delaware was in operation it handled only 61,500 tons. That figure increased annually. By 1850 it topped half a million tons and this was only the beginning. By 1870 the gross tonnage passing through the Deep Cut rose above the million mark. By then the channel had been deepened and widened. But despite the great amount of business it was handling, there was no escaping the grim fact that the Chesapeake and Delaware Canal had cost $2,250,000 to build.

What with the cost of maintenance, repairs and needed improvements there

Delaware City, Delaware. Chesapeake and Delaware Canal.

was no relief in sight for the harried investors. Their only hope was that the United States could be persuaded to purchase the canal and conduct it as a federal waterway.

It would be tedious to dwell on the endless years of debate the proposal produced, both in Congress and without. Numerous committees were appointed to examine the question. In 1919, at the urging of President Woodrow Wilson, Congress acted and the United States bought the canal for a consideration of $2,500,000. In the capable and exacting hands of the U.S. Corps of Engineers it had become a sea-level canal, large and deep enough to accommodate an ocean-going ship plying the Chesapeake or Delaware Bay. These continuing improvements passed the $25,000,000 mark.

Even a brief history of the old canal cannot be concluded without mentioning the service it rendered the government in the opening days of the Civil War. Virginia seceded from the Union on April 17, 1861. Within hours its troops were marching northward to capture Washington. On April 19, the Sixth Massachusetts Regiment, hurrying to the front, was attacked in Baltimore. That night every bridge on the Philadelphia, Wilmington and Baltimore Railroad between Baltimore and the Susquehanna was burned. There remained no way of transporting troops by rail to the capital; Perryville, a small town on the north bank of the mouth of the Susquehanna was as far as they could get.

On April 20 the government seized all steamboats in the vicinity of Philadelphia and dispatched them down the Delaware River, hoping that their shallow draft would permit them to pass through the unfinished canal. Most of them made it and were at Perryville by daylight. As soon as the boats were loaded, they steamed on to Annapolis, from where the troops were able to proceed to the capital by rail. The advance battalions of the Confederacy had reached Washington and were threatening the city, but were turned back. By grace of the Chesapeake and Delaware Canal the boys in blue had arrived at the front in time.[6]

Before leaving the Chesapeake Bay area, the short canal across the wide, marshy estuary of the Susquehanna at Port Deposit, five miles upriver from Perryville, should be noted. Chartered in 1783 to build to tidewater above Havre de Grace, the canal company was financed exclusively by local Maryland money. Waste and bad management kept it from getting anywhere. Hezekiah Niles, editor of the Baltimore *Evening Post*, commented caustically that "more money than water is being poured into the Port Deposit Canal."

By 1800, $150,000 had been spent, and it became apparent that much more was needed to complete the job. The project limped along, with stockholders becoming increasingly wary of sending good money after bad. In 1817 the undertaking collapsed and the canal was never completed. How important and prosperous it could have been was revealed in 1839 when the Susquehanna and Tidewater Canal was built to Havre de Grace and within two years became one of the busiest canals in the United States.

5

The Dixie Canals

A N AUTOMOBILE HAS made the name Cadillac a familiar one to American ears. But our acquaintance with Sieur Antoine de la Mothe Cadillac, the French military commander, largely stops there. Early in the eighteenth century, when France claimed all of this continent lying north of the Great Lakes, he was the king's first officer, entrusted with the security of French Canada, which included the highly strategic post at Detroit. This explains why Cadillac as a place name, is frequently encountered in present-day Detroit and certain parts of Michigan.

Cadillac is of interest to this narrative only because in 1707 he addressed a memorial to King Louis XIV outlining a plan to bypass Niagara Falls and the impassable gorge of the Niagara River by building a short canal to connect Lakes Ontario and Erie. This suggestion undoubtedly reflected the fact that as a Frenchman Cadillac was acquainted with the network of canals by which the internal commerce of France and the Low Countries had moved for centuries. With its numerous locks and aqueducts the famous Languedoc, connecting the Bay of Biscay with the Mediterranean, was the most important canal in all Europe.

The reader is undoubtedly aware that after the passing of more than two centuries such a canal, the Welland, was built, but by the English. France, defeated in the long battle for North America, had retired from this continent. Succeeding generations of Americans have become acquainted with the Welland Canal, if only by name, largely through millions of picture postcards that visitors to the Falls mail home to friends and relatives.

No canal ever conceived had a greater potential than the Welland. Certainly it has been realized, for today as the Welland Ship Canal it is a vital link in the great St. Lawrence Seaway, the joint accomplishment of the United States and Canada, a masterpiece of inland navigation opening the Great Lakes to the water-borne commerce of the world.[1]

Joliet and Cadillac were not the only Frenchmen on this side of the Atlantic to express an interest in canals. In 1796 in Louisiana, then a Spanish possession, Governor Baron François Louis Hector de Carondelet, a Spanish national but of French-Flemish origin, recognized the advantage to New Orleans of a connection by water with Lake Pontchartrain, and ordered the digging of a canal from the ramparts to Bayou St. John, two miles north of the city, which drained into Pontchartrain.

This short canal—the Carondelet—dug with slave labor requisitioned from the neighboring plantations, opened a shorter route to the Gulf than by the Mississippi River. Although only a ditch fifteen feet wide, it proved to be a boon to New Orleans, bringing to the basin outside the ramparts flotillas of bateaux from the far side of the great lake, bearing plantation products and lumber and charcoal. From the Gulf Coast came small schooners with cattle, fish and heavy timber.

Along the spongy, island-dotted Louisiana and Texas coast, several so-called canals came into use. They were largely natural waterways. In stormy weather they were of the utmost importance to fishermen, shrimpers and oystermen.

Not only was the little Carondelet one of the earliest American canals, but also it shares with the Dismal Swamp Canal, which straddles the Virginia-North Carolina line, the distinction of being the longest in continuous service. Both were dug with slave labor. Otherwise they have little in common.

The Dismal Swamp, commonly referred to in those early days as the Great Dismal Swamp, lies, as has been noted, partly in Virginia and partly in North Carolina, embracing an estimated area of twenty-two hundred square miles. Out of it the Pasquotank River flows in a southerly direction and discharges into Albemarle Sound. The promoters of the Dismal Swamp Canal proposed to dig their canal northward from the headwaters of the Pasquotank to a point on Chesapeake Bay, a distance they estimated as a scant twenty-two miles. Several rough surveys were made, no instrument other than a hand compass being employed. The figures arrived at indicated that the land was so flat and so nearly at sea level that it made little difference which course the canal took to reach the bay. That appears to have been true. The little port of Deep Creek, six miles west of Norfolk, was finally designated as the point at which the canal should reach the Chesapeake.

The promoters of the Great Dismal Swamp Land Company (originally known as Adventurers for Draining the Great Dismal Swamp) were interested not only in opening a passage by water to Albemarle Sound that would put the so-called lost provinces of northeastern North Carolina ("lost" because they had no way of sending their farm products and fish to market) in touch with the busy

Virginia ports; they also expected, by drainage, to reclaim great areas of the swamp and harvest its timber.

Save for its stands of bald cypress, black gum, juniper and pine, the swamp was considered valueless. "A refuge for wild animals and reptiles," wrote Colonel William Byrd who was with the expedition that established the boundary line between Virginia and North Carolina. Later visitors, including Thomas Moore, the poet, found it "sublimely beautiful with its tangles of honeysuckle, Virginia creeper, wild grapevines, glossy myrtle, dogwood, mimosa and wild plum."

Although there was very little money in the treasury, construction of the canal began in 1787. The proprietors received an unexpected boost when George Washington, returning home briefly from the Constitutional Convention in Philadelphia, was induced to inspect the Dismal Swamp undertaking. Convinced that it was feasible, he subscribed for $500. Following his lead, wealthy George Mason, James Madison and other prominent Virginians subscribed for small amounts. With the financial situation relieved, at least temporarily, the work continued at a quickened pace.

Washington's interest in the Dismal Swamp Canal deserves being cited, for he was one of the earliest and staunchest supporters of a national canal program, especially as it related to the opening of the West. Although it will be covered later, it should be interjected here that in 1785 he accepted the presidency of the Patowmack Canal Company which had been chartered to build a series of connecting small canals up the Potomac River that, eventually, would result in a through waterway to the Ohio. That same year, he also accepted the honorary presidency of the James River Canal Company.

It should be remembered that aside from the several post roads connecting the centers of population, the country had very few improved roads or highways. If it was to emerge as a viable nation, its pressing need was a system of internal transportation that would end the isolation of villages, small towns and farmsteads. In a rising chorus the public expressed the opinion that a network of canals would solve the problem. Had it been otherwise it is doubtful that the rash of canal building that followed would have occurred. By 1793 no fewer than thirty canal companies had been incorporated in eight of the original thirteen states.[2]

Christopher Roberts, the Harvard economist, in his study of the canal era in New England, stresses the contribution made by public-spirited men in the early days of canal financing. Obviously most early Americans invested in the various canal companies with the expectation of reaping a profit. But there were some men of wealth and high position who risked their capital in the conviction that a system of internal waterways would advance the prosperity of the nation as well as prove to be a sound investment. George Washington was one of them.

Although the Virginia legislature authorized the building of the Dismal Swamp Canal as a public enterprise in 1790, it was built by private subscription, employing slave labor, which, says Alvin Harlow, "was inefficient and often at the point of rebelling at the strange work to which they were driven."[3] Despite delays and annoyances, a section of the canal was put in operation by 1794 but only to flatboats of six-foot beam and drawing no more than two feet of water. By then, contrary to the previously held opinion, it was obvious that several locks would have to be installed to get boats down to the water level of the Pasquotank River. Even wooden locks were expensive to build. With the treasury bare once more,

Great Dismal Swamp Canal, Virginia, which connects Chesapeake Bay with Elizabethtown, North Carolina. *Courtesy United States Department of Interior.*

work stopped. Years passed before it was resumed. By 1807 $100,000 had been expended on the project.

When the canal hooked into the Pasquotank it still was forty miles from Albemarle Sound; but by then most of the trying problems had been conquered. Aside from a few scattered cabins, the only sign of civilization on the length of the river was the shabby, semi-lawless settlement at the mouth of the Pasquotank which began life as Reading and changed its name to Elizabeth City soon after the so-called shingle-getters swarmed in with the opening of the canal to grub out the juniper and oak timber in the great swamp.

The shingle-getters were a wild, reckless, hard-drinking breed that made their own laws as they went along. Like the tie-hacks of Wyoming of a much later day, they brought their feuds and quarrels to town. Many were settled, with blood-letting and cracked heads, in Elizabeth Tooley's tippling house, with the redoubtable Elizabeth acting as referee. It was for her, not for Queen Elizabeth, say local historians, that Elizabeth City was named.

The shingle-getters were skilled in the use of the broad axe and the hand-split juniper shingles that went up the canal by the thousands were reckoned to be the finest produced anywhere. Juniper shingles were not subject to rot and would last until worn thin by the elements. As a consequence, they commanded a high price and were always in demand.

How many bales of shingles went up the Dismal Swamp Canal to Deep Creek, Portsmouth and Norfolk cannot be estimated, but the total must have been tremendous. It was not only shingles that went north via the canal; for many years the Dismal Swamp Canal was the most important means of transportation and communication between the northeastern counties of North Carolina and the ports and markets of Virginia. When the Albemarle and Chesapeake Canal, some miles to the east, came into operation, it took away some of the business. That was sometime after 1828, the year the Dismal Swamp Canal was completed.[4]

Legends endowed the Great Dismal Swamp with a wide assortment of terrors: murderers, fugitives from justice, moonshiners, poisonous snakes and various pestilences, most of which were imaginary. The reputed terrors did not hold back the shingle-getters who made themselves at home in the great swamp. To get what they wanted, they scarred the swamp with fire and axe but they did not tame it.

Although they had neither legal right nor title to the timber or the land, the Great Dismal Swamp Land Company and other lawful owners made no attempt to eject them, being satisfied to have them open up the swamp with their axes and great fires that burned through the slashings and ignited the peat bogs (some of which burned for years), thereby improving the natural drainage and hastening the day when lumbering operations on a big scale could begin. The shingle-getters have long been gone and the great swamp has shrunk to 750 square miles. Sawmills whine where once was heard only the ringing of the axe.

Today, as you glide along what was originally the Dismal Swamp Canal (now an alternate route of the Intracoastal Waterway), you can catch glimpses of the towpath among the trailing vines and flowers that cover it. In the meantime, Elizabeth City at the mouth of the Pasquotank has been for years the thriving trade center of northeastern North Carolina, bearing no resemblance to the rough, tough settlement it was in the days when the shingle-getters had the run of the town and Elizabeth Tooley's notorious groggery was its social center.

Concurrently with the digging of the Dismal Swamp Canal, another canal was under construction 125 miles to the south in adjoining South Carolina. Through no fault of its promoters, it was destined to become one of the most controversial canals ever built. Incorporated as the Santee and Cooper Canal, it was popularly known as simply the Santee.

The little Cooper River, originating in the vicinity of Lake Moultrie, flows southward into Charleston harbor. The Santee, one of the state's most important rivers, cuts across South Carolina from west to east. With its tributaries it drains into the north-central and most productive section of the state. From the Piedmont region, in the west, down through the plantations, the general movement to market of everything being produced was toward the sea. The Santee would have enjoyed almost exclusive control of that waterborne traffic but for the labyrinthine maze of swamps and small islands at its mouth, fifty miles north of Charleston, that made navigation into the ocean almost impossible.

At one point a short distance north of Lake Moultrie, the Santee and Cooper rivers came within some twenty miles of each other. It was reasoned that a canal connecting the two rivers would benefit a major part of the state and increase the prosperity of Charleston. When the idea was first broached it met wide acceptance. The promoters formed themselves into a company and in 1786 the legislature granted it a charter. It was 1792, however, before work on the canal began. Opposition arose as soon as the route was staked out and the rice planters along the coast saw it cutting through their fields. They claimed that the canal would lower the water level and damage the irrigation of their plantations.

At the time, indigo and rice were the principal exports of South Carolina and prosperity was measured by the extent of their annual production. Indigo plants were introduced into South Carolina from the West Indies by Eliza Lucas, who later married the celebrated Charles Pinckney.[5] Cultivating indigo became a sizable industry, but rice was the mainstay of South Carolina's economy. In 1860 the state produced three and a half million bushels, which was better than half the total amount of rice produced in the United States. But that is an era long since gone and today it is difficult to find anyone who can even remember the names of the great plantations that once fringed the Waccamaw and other little tidal rivers, each with its own private landing and manor house. They are all gone, deserted and destroyed, with only the decaying chimney of a rice mill to be glimpsed occasionally through the encroaching sub-tropical greenery.

What few people realized at the end of the eighteenth century was that the economy of the state was changing, and the changes were not to be of a temporary nature. Prior to 1790 cotton had not been grown successfully in South Carolina. Without fanfare, new varieties were introduced for which the soil and climatic conditions of certain parts of the state proved to be ideal. There was a gradual moving away from the coast.

Its immediate effect on the Santee and Cooper Canal was to increase the cost of building it. Slave labor had been cheap; now slave owners found it advantageous to keep their slaves working cotton. By 1795, with only five miles of the ditch completed and several wooden locks installed, the company found itself out of funds. The directors appealed to the state for aid and were permitted to conduct a lottery, the favorite device of financially embarrassed canal companies. Between the money so derived and numerous assessments on the stockholders, the canal was completed in 1800 one of the first canals to be put in operation. Despite the ignorance of John Senf, the state engineer who built it, and his unwillingness to seek or accept advice, it was basically well built, and after several artificial reservoirs improved its water supply, its success was assured.

The Santee was a long time paying off its indebtedness. But eventually it did. Ironically, a good part of the tonnage the big flats brought down to Charleston harbor was baled cotton. In the first weeks of the War Between the States, when the attention of North and South was focused on Charleston, the Santee was used to advantage by General Beauregard, the Confederate commander, for moving troops.

Long before the war, a dusty and forgotten bit of history occurred on the Santee that is worth recalling. You will remember that John Sullivan, after severing his connection with the Middlesex Canal, engaged in a long and bitter legal battle with the monopolists backing Robert Fulton concerning ownership of his invention, the steam tug. To raise money to continue the fight, he brought his tug, the *Cygnet,* to Charleston and proceeded to demonstrate it on the Santee. It performed so well that he was able to sell the little vessel to a group of Georgians who put it on the Savannah River where, unfortunately, its boiler exploded and it sank.

6

The Chesapeake
and
Ohio Canal

I N THE YEARS before man defiled it the Potomac River came tumbling bright
and clear out of its sources in the Alleghenies beyond Cumberland, and by twisting
and turning fought its way down to tidewater and Chesapeake Bay, a distance of
185 miles. The river is the boundary line between Maryland and Virginia. The
numerous loops and turnings of that line reveal how one obstacle after another
turned the river back on itself in its tortuous journey to the sea.

Save for short stretches the Potomac was unnavigable. The Great Falls and
Little Falls above Georgetown were impassable impediments. And yet many men
of substance believed that by a series of short canals around the worst obstacles
to navigation the water could be opened to Cumberland and a way found from
there to connect the Potomac with the Ohio River. The first and ablest proponent
of such an undertaking was George Washington.

In 1774, prior to the Revolution, he had introduced a bill in the Virginia
House of Burgesses for the improvement of the navigation of the Potomac. It was
opposed by the electors from the central and southern part of the colony on the
ground that it would not benefit their constituencies. He then amended his bill to
include improvement of the James River. In that form it might have been passed,
but the war was at hand and all talk of such improvements ended abruptly. But
through the trying years of military duty that followed, Washington's interest in

One of the original locks on the Chesapeake and Ohio Canal at Georgetown. *Courtesy National Archives.*

Old towpath of Chesapeake and Ohio Canal still discernible in the weeds. *Courtesy New York Public Library Picture Collection.*

opening a trade route to the Ohio country remained constant. In his personal correspondence there are repeated references to it.

No American of his time was as well acquainted with the Appalachian Mountain wilderness as he. As a surveyor in 1748 at the age of sixteen, on horseback and on foot, he had first made his way through its passes and defiles to establish the boundaries of the great Fairfax estate. When a group of Virginians acquired an immense tract of land in southern Ohio in 1749, and organized a trading concern doing business under the title of the Ohio Company, they blazed a trail up the Potomac and down the Monongahela to reach their holdings, which they had just been granted. Washington helped to survey a road over that route the following year. Three years later he followed it at least part of the way as Governor Dinwiddie's envoy to the French commander at Fort LeBoeuf.

He traveled in this instance as Major George Washington, adjutant general of the Virginia militia. The letter he carried called on the French to remove themselves from territory that obviously belonged to the king of England. Legardeur de Saint-Pierre, the French commander, ignored Dinwiddie's demand and proceeded to establish himself at Fort Duquesne, where Pittsburgh now stands. War followed, attack and counterattack. When General Braddock, with English regulars and Virginia militia, made his ill-fated march on Fort Duquesne in 1755, Washington accompanied him and suffered the humiliation of defeat.

View from Round Top Mountain, Hancock, Maryland, showing the railroad, canal and Potomac. *Courtesy National Archives.*

In the three years between his mission to Fort LeBoeuf and the Braddock disaster he had made three long wilderness journeys to the head of the Potomac and the James rivers. What he learned had convinced him anew that a water route could be built to the Ohio.

With peace and the independence of the United States secured by the Treaty of Paris in 1783, Washington returned to Mount Vernon and began an intensive campaign by mail to drum up support for legislation designed to open communication with what was commonly referred to as the Ohio country, either by government road or canal. "The great object," he wrote to Edmund Randolph, Virginia's attorney general, "for which I wish to see the navigation of the Rivers James and Patowmack extended is to connect the Western Territory with the Atlantic States; all others with me are secondary."

Washington found an influential ally in Governor Benjamin Harrison. Harrison told him that if he would make still another transmountain journey into the Appalachian wilderness and submit a detailed report of what he found, he (Harrison) would present it to the Assembly with his recommendation.

On September 1, 1784, Washington left Mount Vernon, accompanied only by a "trusted servant," [1] on his historic tour of the mountain barrier beyond the headwaters of the Potomac. He explored the south fork of the Shenandoah, crossed over to the Kanawha and down the Monongahela to its conjunction with the Allegheny, where the united rivers become the Ohio. Of necessity, this long journey —estimated at 650 miles—had to be made on horseback and on foot.

Even then George Washington was being hailed as the Father of His Country. Certainly he was the most famous and respected man in the United States. One can only wonder what the reaction of the wilderness dwellers—he noted in his journal how numerous they had become—must have been on learning his identity.

The long detailed report he tendered Governor Harrison dealt not only with the immediate needs of the country to the west but also with the future needs of the whole Northwest Territory. The governor so strongly approved it that action by the Assembly was immediate. Opening the Potomac and the James became the first order of business. In May, 1785, a bill incorporating the Patowmack Company was passed; three months later the James River Company was chartered.

Washington was a prime mover in organizing the Patowmack (the original spelling of Potomac) Company, and as noted, was elected company president. So much enthusiasm developed at its first meeting that more than four hundred shares of stock with a par value of $400 each were subscribed.

From Fairfax Stone, above Cumberland, Maryland, where the Potomac rises to Point Lookout below Washington, D. C.—which did not yet exist—the distance was 184 miles. It was not the immediate concern of the Patowmack Company to build a canal from Cumberland to salt water; what it proposed to do was to build a series of short canals around the worst barriers to navigation and endeavor to connect them only when they had proven profitable enough to justify the expense.

In recognition of his services the directors of the Patowmack Company made Washington a gift of fifty shares of stock. He accepted them with the proviso that he be permitted to donate them to a university of his choice. When the Virginia legislature several months later presented him with a hundred shares of James River Company stock, he bequeathed it to what eventually became Washington and Lee University at Lexington. Apart from this gifting, he invested his own funds

George Washington Patowmack Canal channel at Great Falls, Virginia. *Courtesy National Archives.*

heavily in both companies. This was two years prior to his becoming an investor in the Dismal Swamp Canal Company.

In the spring of 1786 two hundred laborers, indentured immigrants and slaves, were set to work on the Patowmack Company's undertaking at the Great Falls above Georgetown. Harlow calls it "the first corporate work in America on an improvement of navigation for public use." [2]

The Potomac was navigable for ocean-going vessels as far upstream as the Georgetown Basin at the foot of Rock Creek. Alexandria was the most important town on the river, but in the fall, when the big, heavy wagons of the farmers of the Potomac Valley rumbled down the so-called wheat roads to tidewater with grain, Georgetown Basin was thronged with shipping, there to transport cheap Virginia and Maryland grain to Europe. Alexandria still retained the prestige and prosperity it had won in the years when tobacco was the only crop produced in eastern Virginia. So completely had tobacco dominated trade in those days that it was often used as currency. But putting the land to that one crop, year after year, had exhausted the soil, and no longer was the tobacco market at Alexandria the lively place it had been.

The Chesapeake and Ohio Canal cut across Rock Creek Park, at Washington, to reach the Potomac. *Courtesy National Archives.*

Another view of the original Patowmack Canal around the Great Falls, Virginia. *Courtesy National Archives.*

Harpers Ferry, 1859.

The Potomac Valley has been called the Battleground of the Civil War, for it was there that the contending armies of the North and South dueled from the first Battle of Bull Run to Lee's retreat from Gettysburg. But that conflict was far in the future when the Patowmack Company began work at Great Falls. Above Georgetown there were only two established towns of any size, Hagerstown and Cumberland. In addition, there were a number of small settlements, centers of farming communities. For a market, however, they looked to Baltimore rather than to the lower Potomac.

For years pioneer Scottish and German settlers had been pushing out from the Susquehanna and settling in northern Maryland. They were seeking religious freedom as well as economic security. Being of many different sects, they separated into distinct communities.

Back in 1734 Robert Harper, a Scottish immigrant, established the ferry just above the juncture of the Shenandoah River and the Potomac which John Brown, the abolitionist leader, was to make famous more than a century later. There the great natural routes of travel ran north and south through the wide limestone valleys of the Blue Ridge Mountains.

At Shepherdstown, a few miles above Harpers Ferry, James Rumsey, one of many claimants to the title of inventor of the steamboat, had established himself.

He was an educated man, recognized widely as an engineer as well as an inventor. He had never seen a canal or canal lock when, unable to find a better qualified man, the Patowmack Company put him in charge of operations.

New and vexing problems faced Rumsey from the day he took charge. He soon discarded the idea of using slack-water navigation to get around the falls; a canal would have to be blasted through solid rock and a series of stone locks installed. To speed the work he devised the scheme of working there in high water and moving his crew upriver to other locations in seasons of low water. He had upwards of a thousand men on the payroll, many of them Irish lately arrived in the States. They fought among themselves, were rebellious and often drunk, for in addition to the company ration of three gills of rum a day, they found other sources. The indentured laborers ran away. Those who were returned by deputy sheriffs had their heads and eyebrows shaved as punishment, but at the daily roll call men were always reported missing.

Canal tunnel on the Chesapeake and Ohio Canal, c. 1890. *Courtesy Edwin P. Alexander Collection.*

Before he finished, Rumsey built five canals, the longest—3,814 feet—at the Little Falls. His greatest achievement was the canal and the five locks at the Great Falls. When completed in 1802, it was hailed as another of the great engineering achievements of the century. Washington had not lived to see it.

The Patowmack Company's short canals proved to be of great assistance to flats and arks drawing not more than one foot of water. But this was a far cry from Washington's dream of connecting the Potomac with the Ohio. However, the cost of freight by water was reduced to half the cost by wagon. Traffic on the river increased so rapidly that a severe shortage of boats occurred. A total of forty-five thousand barrels of flour, together with much whisky, iron, wheat and tobacco, was locked through, the tolls amounting to somewhat more than $10,000. The Patowmack Company declared its first dividend. Unhappily, it was also its last.

In the meantime the government of the United States had settled in its new home on the Potomac. First known as Federal City, then Washington City, it finally blossomed forth as Washington, D.C. In the beginning the unfinished capital must have been a nightmare for the 131 government clerks and their families who had been torn out of their comfortable Philadelphia homes and shipped to what they referred to contemptuously as "this place in the woods." Abigail Adams, wife of President John Adams, wrote home witheringly, "Pennsylvania Avenue, running from the President's house to the capitol, is merely a rutted six-foot lane through the forest." [3] And again: "Two articles we are much distressed for; one is bells, but the more important one is wood. Yet you can not see wood for trees."

Thomas Johnson of Frederick, Maryland, Washington's boyhood friend, succeeded him as president of the Patowmack Company. Long a substantial shareholder, he was dedicated to the idea of connecting the Potomac with the Ohio. The company charter called for certain costly improvements on the upper river. Although Johnson might have preferred not to proceed with them at once, he did not hold back. He was aware, as not too many others were, of the enormous potentialities of the great bituminous coal beds around Cumberland. Obviously, if the Patowmack Company could open the river to Cumberland, millions of tons of coal would be freighted down to tidewater.

Johnson was not concerned about the progress being made in building the great National Road—a hard-surfaced eight-lane highway that followed the Cumberland Pike from Frederick to Cumberland, from where by way of Uniontown, Pennsylvania, and Washington, West Virginia, it was to climb over the mountains to Wheeling, on the Ohio, across river from Zane's Trace. A road, no matter how costly to build, would be limited to wagon traffic. [4]

He also knew that Baltimore was toying with the idea of experimenting with a railroad. It had been demonstrated that a train of cars, propelled by what was known as a "locomotive engine," could be drawn across country on rails if the rails were laid on level ground. There was very little level ground between Cumberland and Baltimore. Johnson took comfort from that and turned his attention to more pressing problems.

But dark days were ahead for the Patowmack Company. So much time and money were consumed in blasting a channel through the rocky gorge south of Point of Rocks, twenty miles below Harpers Ferry, where the Catoctin Mountains squeezed the river into a narrow defile, that company funds were exhausted. Appeals were made to the legislatures of Maryland and Virginia for assistance.

One of the series of locks by which the Chesapeake and Ohio Canal got around the Great Falls at Georgetown. *Courtesy National Archives.*

Original lockhouse of the Chesapeake and Ohio Canal at Seventeenth Street and Constitution Avenue in Washington, D.C. *Courtesy Library of Congress.*

Below Hancock, on the Chesapeake and Ohio Canal. *Courtesy Library of Congress.*

They went unanswered. A new stock issue was offered for sale but there were no buyers.

These facts were not unrelated, and it soon became apparent that the purpose was to put the old company out of business. In 1821 a joint commission appointed by Maryland and Virginia was named to inquire into the affairs of the Patowmack Company, which, it was charged, had failed to comply with the terms of its charter and seemingly never would. In due course the commission reported that it found the company heavily in debt and likely to remain so; that further aid to it appeared unwise; that it would be sound procedure to cancel its charter and "adopt a more effectual means of improving the navigation of the Potomac."

The Patowmack Company signified its willingness to surrender its charter on liberal terms and bow out. Virginia immediately chartered a new company to be administered jointly by Maryland, the District of Columbia and itself. Maryland rejected the idea, but several months later sent out invitations to Virginia, the District and several southern Pennsylvania counties to join with her in a

Stone bridge over the Chesapeake and Ohio Canal at Georgetown. *Courtesy Library of Congress.*

Pennifield's Lock on the Chesapeake and Ohio Canal. *Courtesy Library of Congress.*

"canal convention" to be convened at the Capitol in Washington in November, at which ways and means of effecting a connection between the Potomac and the Ohio could be considered.

The response was excellent, and in mid-November, 1823, twenty-six delegates were present. Some years back General Simon Bernard, the famous French engineer, had rendered to the Patowmack Company an estimate of the cost of constructing a canal from Georgetown to Pittsburgh. He had placed it at $22,000,000. For the past months James Geddes and Nathan Roberts, Erie Canal engineers, had been going over Bernard's report and they expressed the opinion that his figures were excessively high. When they submitted their revised report to the delegates the latter were aghast, for although the two men had trimmed the Frenchman's figures by many millions of dollars, they pegged the cost at $11,000,000.

No further talk was heard of building across the mountains to Pittsburgh. A far more pressing problem was to open a canal from Georgetown to Cumberland. Geddes told the assemblage that by using the extensive work done at the Great Falls and incorporating some of the other improvements the Patowmack Company had installed the total cost of construction would not be over $4,000,000.

This ended further unrealistic talk of digging a waterway through the mountains to Pittsburgh, and the delegates addressed themselves to the question that had really brought them together. It was not whether a canal extending from Georgetown to Cumberland, taking its water from the Potomac but otherwise independent of the river, should be constructed. They were agreed on that. But there was no agreement on who was to build it and who was to pay for it.

The federal government further complicated matters by insisting that the contemplated canal should be sixty feet wide at water surface and six feet deep, which would materially increase Geddes' estimated cost. In return, however, the government pledged itself to invest $1,000,000 in the project. It was a compelling argument.

Before adjourning, the convention recommended a subscription of $2,750,000 to the capital stock of the proposed company that was to build the canal, equitably apportioned to Virginia, Maryland, the District and the federal government, with total capitalization not to exceed $6,000,000.

Before organization of the new company was complete, Virginia rechartered it as the Chesapeake and Ohio Canal Company. This headlong start was not indicative of the snail-like pace to which it was reduced while finding subscribers to its stock. It was not until June, 1828, that sufficient subscriptions had been secured to warrant accepting the Virginia charter. At that time, after five years of solicitation, only $3,608,900 worth of stock had been disposed of: the federal government had taken $1,000,000, the state of Maryland $500,000, the city of Washington $1,000,000, and Georgetown and Alexandria $250,000 each, and the balance sold on the open market. Baltimore contributed nothing. Having discovered that its pet scheme of a branch canal from Point of Rocks to Baltimore, which would have made that city the principal terminus of the Chesapeake and Ohio Canal, was financially impracticable, it had decided to link its future to the development of the railroad. Virginia, too, had lost interest in the canal when the surveys showed it going up the Maryland side of the Potomac all the way to Cumberland, and contributed nothing.

Locking through at Hancock, Maryland, on the Chesapeake and Ohio Canal. *Courtesy Library of Congress.*

An old lock on the Chesapeake and Ohio Canal at Hancock.

At Georgetown, on July 4, 1828, the established day for the launching of such undertakings, with flags, cannon booming, a military band enlivening the occasion with martial music and John Quincy Adams, president of the United States, on hand to turn the first spadeful of earth as "a great concourse of ladies and gentlemen cheered," the work of building the Chesapeake and Ohio Canal began. So slowly did it proceed, so many were the stoppages, that the young men and women who cheered President Adams so lustily on that historic occasion were middle-aged by the time water was let into the canal at Cumberland twenty-two years later.

Perhaps it was symbolic of the difficulties the C. & O. was to face that President Adams had to remove his coat and settle to his task before he could drive his spade through a tangle of obstinate roots. There was more significance in the fact that forty miles away on that hot July morning, in Baltimore City, the venerable Charles Carroll, the only surviving signer of the Declaration of Independence, had stepped out of retirement to lay the first stone of the Baltimore and Ohio Railroad.

Even when viewed from the advantage of hindsight there does not appear to have been any reason to believe that in only a decade or two railroads would be offering a challenge that other forms of transportation could not meet. For almost half a century America had been on the move, heading West. The exodus from New England alone was estimated at 400,000. Countless veterans of the Revolution had piled their families and a few belongings into wagons and deserted their rocky hillside farms, often without having any definite destination in mind but

The Chesapeake and Ohio Canal paralleling the Potomac. *Courtesy Library of Congress.*

determined to find a more fertile land and a new start in life somewhere in that vast country they identified vaguely as "the West." Soon, only a caved-in cellar or crumbling blackened chimney remained to indicate where there once stood a farmstead.

Massachusetts and Connecticut towns were similarly deserted. Mills stood idle because the workers had been organized into so-called colonization companies by land speculators from New York State and had "pulled out," as the saying had it, for the new land of opportunity beyond the Hudson. For eight months a year the completed National Road was so snarled with westward-bound traffic that if often took hours to untangle the wagons, the brawling teamsters laying-to with their fists and whips.

The population of Ohio had tripled in the twenty-five years that had passed since it was admitted to the Union as a state in 1803. It now approximated 100,000. Reaching that booming market by canal was an alluring prospect, but the directors of the C. & O. were too harassed by their difficulties at home to give it any consideration. The company was out of funds and in debt for more than $7,000,000. The state bonds it held were going begging and could be disposed of only at a sharp discount. The financial problems of the C. & O. were largely those that beset other canal companies and need not be gone into here. An exception was the banding together of the landowners along the surveyed route who refused to grant the company right of way at a reasonable price, thus compelling the C. & O. to condemn almost the complete route from Georgetown to Point of Rocks and purchase at an exorbitant price more than three thousand acres of land. The C. & O. was dealt another unexpected blow when the cholera epidemic that swept the country in 1832 felled so many of its workers that operations were brought to a long and painful stop.

At the canal convention in Washington in 1823 Alexandria had expressed its interest in the Chesapeake and Ohio Canal by subscribing for $250,000 of its stock. It felt that to guarantee its future prosperity it had to be reached by the C. & O. The only way that could be done was by building an aqueduct across the Potomac at Georgetown and an extension of the canal along the south shore of the river. The canal company had refused to engage in such a costly undertaking, but in 1830, with news from the North of the great success the completed Erie Canal was enjoying spurring it on, Alexandria organized a corporation and secured a charter to do the work itself. The federal government felt compelled to aid the enterprise with a subscription of $250,000.[5]

The Alexandria Company's aqueduct across the Potomac was to be of historic importance in the years to come. It was sturdily built of wood resting on massive stone pillars, and was eleven hundred feet long. Work on it began in 1830 and was completed in 1833. Located where the Potomac was almost half a mile wide, it was then, and for many years thereafter, the only connection between the north and south shores of the river.

After serving the purpose for which it had been constructed for eighteen years, Virginia and the South had cause to regret that it ever was built, for in the opening days of the great conflict between the North and the Confederacy, the United States Army seized the aqueduct, drained off the water and converted it into a foot and wagon bridge. If you have wondered about the origin of the often-mentioned Aqueduct Bridge, this was it. Over it untrained Union soldiers marched

Excursion party on the Chesapeake and Ohio
Canal. *Courtesy Edwin P. Alexander Collection.*

The aqueduct bridge at Georgetown on the Chesapeake and Ohio Canal. *Courtesy National Archives.*

to and retreated from the disastrous defeat at Bull Run. Across it in the years that followed, thousands of men in blue, cannon and munitions of war reached the front. When returned to the owners after the conclusion of the war it was in such deplorable condition that it would not hold water and had to be abandoned.

So many things had conspired to delay the C. & O.—including the long and costly aqueduct over the Monocacy River—that when it finally reached Point of Rocks it found the Baltimore and Ohio Railroad already there. It was a primitive single-track affair, running on light cast-iron rails. But it was indicative of things to come. Without much difficulty it had built up the Patapsco River Valley and crossed over to the Potomac. Its immediate objective was Cumberland.[6]

The lower sections of the canal had been open for some time and were bringing in a surprisingly large amount of tolls. Between tolls, water rents and interest on the bonds it was holding it had managed to continue. But it faced a showdown at Point of Rocks, for there the narrow shelf of land between the cliffs on the opposite side of the river was not wide enough to permit the passage of both the canal and the railroad. The latter secured a right of way. The C. & O. appealed to the courts,

Trees tumbling the ancient masonry of old Lock 2, on the George Washington Canal at Great Falls, Virginia. *Courtesy National Archives.*

The Chesapeake and Ohio Canal aqueduct, across the Monocacy River.

Stone gatehouse on Chesapeake and Ohio Canal, 1855. *Courtesy National Archives.*

Coal boat bound down the Chesapeake and Ohio at Cumberland. *Courtesy Library of Congress.*

claiming prior rights, and after a long legal battle was the victor. The Baltimore and Ohio, not to be stopped, drove a tunnel through the mountain at Point of Rocks and railroad and canal proceeded up the river side by side. The former under court order was compelled to erect a fence between its track and the towpath.

The term "ditch" as applied to a canal was usually used in a demeaning sense. With a channel sixty feet wide and a depth of six feet, the C. & O. could not be called a ditch. It was attracting a population of its own; families who lived the year round on their boats, father doing the steering, mother the cooking and washing, the son (or daughter) on the towpath doing the driving. Small boats that could carry up to twenty-five tons were to be found in almost every side-cut, waiting to pick up a mixed cargo of wheat, corn, barley or potash.

There was a natural enmity between boaters and train crews. After the railroad and canal reached Cumberland, both sides went armed with a supply of rocks to be hurled at the enemy in passing, with the engineer letting go with a blast of the whistle to stampede mules and horses.

The Chesapeake and Ohio Canal reached Cumberland on June 11, 1850. With the entire population on hand to cheer the momentous occasion, water was turned into the canal on October 1. A new boat, appropriately named *Cumberland*, was launched and began the long journey down the waterway. Although the work was considered to be complete, improvements remained to be made. The most notable was the tunnel, 3,118 feet long, that was cut across a five-mile bend in the river. Before entering it, boats lighted their head lamps and drivers carried lanterns.

Having been compelled to buy land on both sides of its channel, the canal company had few bridges to build, which was a great saving. Periodic floods closed navigation at times for several weeks, but the C. & O. had some golden years when tolls topped $1,000,000, most of it earned by the vast amount of coal it was forwarding. Bigger and bigger boats, able to carry up to a hundred tons, became the rule rather than the exception. A line of fast packet boats from Williamsport (canal port for Hagerstown a few miles to the north) to Harpers Ferry and Georgetown was in triweekly operation.

By the late seventies the Chesapeake and Ohio Canal had passed the peak of its prosperity, and the downward slide that began in 1881 could not be checked. The explanation was not hard to find. With heavier rails, bigger rolling stock and more powerful locomotives, the once despised Baltimore and Ohio Railroad was draining the coal-carrying business away from the canal. In 1888 the C. & O.'s net earnings dropped to $2,700. The great flood of 1889 so wrecked it from end to end that it could not continue. New hands tried to put the broken pieces together, but the Chesapeake and Ohio, one of America's most beautiful canals, could not be saved. In April, 1924, it surrendered its charter and suffered the ignominy of having its assets purchased by the Western Maryland Railroad.

It has not been forgotten. Perhaps no other American canal, including the great Erie, is recalled with such affection. Today, in places, it serves as a recreation area. Scores of hikers from all over the country come to trudge along its old towpath and have made it the most popular walking tour in the eastern United States. If you prefer riding to walking, you will find at the Thirtieth Street overpass in Georgetown, where the canal once made its way through town, an ancient barge and a patient mule that will take you several miles up the canal and get you back in time for an early dinner.

Lockmaster's home on the Chesapeake and Ohio Canal.

Chesapeake and Ohio Canal competing with the railroad, Point of Rocks.

In the days when the Chesapeake and Ohio Canal was taxed to capacity. *Courtesy Library of Congress.*

7

The James River
and
Kanawha Canal

BECAUSE HISTORIANS HAVE always written of it that way, we are inclined to think of Tidewater Virginia of colonial and post-colonial days as a land of elegance peopled only by a wealthy aristocracy and its black servitors. Corroborative evidence that the so-called First Families of Virginia—the Byrds, Randolphs, Lees and many others—lived in splendor is to be found in the great manorial homes that graced the plantations along the James below Richmond.

Upriver to the west and in the Great Valley of Virginia there was, as yet, no reflection of this magnificence. The region was too young to have become fettered with social traditions. Grandfathers of living men had known it when great herds of buffalo passed through on their annual migrations back and forth from north to south, and when Indian war parties crossed the James at the ford above Looney's Creek, coming and going in their wars with the Cherokees and Catawbas. Its towns were still small, but its prosperity awaited only a practical and economic means of getting its wheat, barley, flax, smoked hams, barreled pork, tobacco and whisky (the last of a quality that was to make it famous) to market.

As previously noted, Virginia enacted legislation in 1785 calling for the improvement of navigation of the James River. In August of that year the James River Company was organized. The charter specified that the company was entitled to collect tolls if it kept navigation open in the dry seasons for boats "drawing at least one foot of water." It was further obligated to build a short canal around the falls (at Richmond). Nothing was said about removing or circumventing other known barriers to navigation, it being taken for granted that in its own self-interest the company would open the James as far upriver as possible.

For the value of his name in disposing of its stock, the promoters endeavored to have George Washington accept the presidency of the company. The demands on his time were such that he could not accept, but he compromised by agreeing to act as honorary president under Edmund Randolph, its active head.

The James crossed Virginia from west to east and was the state's most important river; it pursued a serpentine course that came close to doubling its length. Up beyond Buchanan, at the southern tip of the great central Valley of Virginia, where it began, the mountains squeezed the Cowpasture and Jackson rivers together. Two miles below Iron Gate they united to form the James. There, with the Alleghenies to the west and the gentler Blue Ridge Mountains to the east, it began its three-hundred-mile journey to tidewater and the sea.

The first work undertaken by the company was digging a canal from Richmond around the falls to Westham, seven miles upriver. Although only a channel thirty feet wide and three feet deep, the work dragged on for years and was not completed until December 29, 1789, when the company received a loan from the state. Open passage around the falls having been secured, the effect on business was immediate. Richmond felt it almost at once. Never before had so many big flats, laden with flour, whisky, hogsheads of tobacco and hard grains, been poled down the James. Many of them came from as far as Buchanan, 196½ miles upriver from Richmond. With the removal of the chief obstacle to getting their produce to market by water, planters in the lower Piedmont hurriedly began building boats to take advantage of the cheap transportation. It cost very little to construct a big flat. It could be loaded to the waterline, and a crew of three husky slaves was all that was needed to get it down to Richmond—two men handling the poles and a third doing the steering. They made excellent rivermen and were proud of their dexterity with the shod poles and the sweeps.[1]

It was a different story at Balcony Falls, thirty miles above Lynchburg, where the river cut a gorge a thousand feet deep through the Blue Ridge Mountains and the water fell two hundred feet in four miles. Boats were lost and crewmen drowned in running these dangerous rapids. Those that got through rarely attempted to return. It was the usual practice to sell the boats in Richmond for what they would bring as secondhand lumber. But flats could be poled back to Lynchburg, 146 miles above Richmond. All that was required was time and muscle, the return usually taking as long as ten days.

Perched on the north bank, Richmond looked down on the boiling, rocky, island-studded rapids of the James and beside it the placid little canal that had been dug along the left bank. There was an even broader view to the east, where the James became a tawny, navigable tidal estuary, taking its color from the land through which it flowed. Being a commercial city, its prosperity linked to the river, Richmond demanded its further improvement.

Under pressure, the James River Company kept a large work force employed on the upper river between Lynchburg and Crow's Ferry from 1796 to 1801. Black men, rented slaves, did the laboring. Former field hands, they soon developed a fondness for the river, which found expression in the campfire songs they originated. Several became heroes. Just below Balcony Falls within a few feet of what today is the roadbed of the Chesapeake and Ohio Railroad and was in former times the towpath of the James River Canal, there is a granite monument erected to the memory of Frank Padget, "a colored slave who during a freshet in James River in 1854 ventured and lost his life by drowning in the noble effort to save some of his fellow creatures who were in the midst of the flood." Five men drowned that day when the canal freighter *Clinton* was destroyed.

In seasons of drouth the James River Company had not been able to maintain the depth of water required under its charter. It was obvious by now that this could be done only by installing a number of locks and dredging the channel. Rather than undertake that costly improvement, it sought to escape criticism by declaring a 5 percent dividend on its capital stock, of which there was $210,000 outstanding. It brought the stock up to par, but instead of diminishing criticism, this action had the reverse effect: the charge was widespread that the James River Company was more interested in enriching its stockholders than in improving the James.

The storm continued. It was charged that while the company was declaring annual dividends of 15 to 20 percent and collecting its exorbitant tolls, it was allowing the improvements it had made to deteriorate to the point where boats were continually grounding for lack of water, with their cargoes spoiling and becoming unsalable. In 1812 a legislative committee was appointed to look into the matter. So here, for the first and only time in the history of American canals, a company found itself in financial difficulties not for lack of funds, the usual situation, but because it was too prosperous.

In its report the committee found the improvement of the river far behind schedule but, in view of the difficulties the company had encountered, held it blameless. In effect, it was a political whitewash and was so received by critics of the James River Company. As a sop to the dissidents, a board of commissioners was appointed by the legislature in 1812 to report on the feasibility of Washington's original plan to connect the James with the Kanawha and the Ohio. The commissioners left the James at Buchanan, crossed the mountains to the Greenbriar, went down that stream to New River and on to the Kanawha and the Ohio. The report they rendered was so favorable that unexpected support for the undertaking developed in the legislature. When the James River Company was asked if it would build such a waterway if adequate tolls were guaranteed, it rejected the invitation.

The company's unwillingness to participate in the proposed venture was taken as evidence that some sort of secret agreement existed between it and certain political leaders. The state owned a majority of the James River stock and was—presumably—collecting a handsome share of the dividends. Naturally officials were reluctant to tamper with this lucrative state of affairs.

The clamor continued but for three years the state did nothing to quell it. In the meantime the James River Company did considerable work improving navigation on North River, one of the principal tributaries of the James which joins it just above Balcony Falls. Had North River been dammed and a suitable reservoir established, it would have solved the water problem on the lower James. The best reason the company had for not doing so was that twenty miles up North River was the important town of Lexington, the center of a rich agricultural region which it hoped to connect with James River navigation.

Finally, under pressure, the state brought an action in Superior Court to

Balcony Falls on the James River and Kanawha Canal, which parallels the James.

vacate the charter of the corporation. No matter how well intentioned it may have been, it resulted in an arrangement that was worse than ridiculous. Perhaps it was a necessary forerunner to the sound, businesslike improvement of the James which eventually occurred.

While the case against the corporation was still pending, it voluntarily surrendered its charter. The action against it was dismissed and the state took over. Under the new ownership, with confidence restored, James River Company stock climbed to a new high. The legislature responded handsomely and, with the openhandedness of politicians with the people's money, guaranteed stockholders an annual dividend of 12 percent up to 1832 and 15 percent thereafter.

Presumably, if the required dividends were not earned, the state would make up the deficit. This bit of folly was compounded by engaging the defunct James River Company to proceed with the river work as the state's agent.

Led by the Richmond *Whig*, so much public indignation was aroused that in 1823 the company informed the state of its inability to proceed and gave up the job. The legislature named a board of commissioners to carry on. Instead of regarding their appointment as a political sinecure, they bent to their task and did a surprisingly good job. On their recommendation Virginia authorized the building of a short canal around Balcony Falls and the extension of the Richmond canal from Westham to Maiden's Adventure, twenty-seven miles upriver, in 1824. The most important thing they did was to engage Benjamin Wright, of Erie Canal fame, to "advise, conduct new surveys and render estimates of the cost of projected improvements."

Wright's estimate of the cost of the extension to Maiden's Adventure was so much higher than expected that the legislature refused to go forward with the work. Instead it confined itself to building the canal at Balcony Falls and constructing a turnpike over the crest of the Alleghenies.

The two-hundred-mile Allegheny Turnpike was not a controversial matter; it had been considered even before there was talk of improving the James. Work was begun that year and was finished in 1828.[2]

Not all of the old crossings of the upper James have survived the march of time. Wherever a man operated a ferry, the crossing took his name—Looney's Ferry, Crow's Ferry, Lynch's Ferry. Lynchburg was known as Lynch's Ferry in the first years after John Lynch established his ferry there and built the first tobacco warehouse on the upper James. Having located in the heart of the South's dark tobacco country, he prospered. So did Lynchburg. In 1828, second in size only to Richmond of all the James River towns, it demanded the calling of a state-wide convention to end the dawdling over improvements and decide whether or not the complete canalization of the river should be undertaken.

It was not until May, 1828, that such a convention was authorized. It met at Charlottesville. Attending as delegates were many noted Virginians, including John Marshall, Mason, Madison, Monroe. The delegates who were convinced that a canal should be dug up the James and a connection made with the Kanawha found an eloquent spokesman in Justice Marshall. "If Virginia did nothing to claim its rightful share of the ever-increasing western trade, New York and Pennsylvania, with their system of canals, would soon monopolize it."

It was a persuasive argument. After three days of debate the convention made public its conclusions, which were to construct a canal up the James to Buchanan and from there "by the best practicable route" make connection by water with the Kanawha. It further recommended that a private corporation be chartered to do the work. On receiving the report the only action the legislature took was to turn it over to a committee, and forget it.

John Marshall refused to let the matter die, and in 1832 he succeeded in chartering the James River and Kanawha Company. But it was not until 1835, the year of his death, that the new company was able to organize. The Federalists were again in power. As state leader of his party, Marshall succeeded in getting the legislature to turn the canal project over to the James River and Kanawha Company. It was capitalized for $5,000,000, Virginia subscribing for three-fifths of that amount; two of the state's leading banks took $500,000 each; and the cities of Richmond and Lynchburg bought small amounts. There was a string attached to this agreement, and a very tough one it was: the James River and Kanawha Company agreed to credit the state $1,000,000 for the equipment—the former property of the old James River Company—that was turned over to it. The equipment proved to be of so little value that the $1,000,000 had to be written off as an almost total loss.

With Benjamin Wright reengaged as consulting engineer, and men who had learned their trade on the New York State canals having been brought down from the North, the James River and Kanawha Company went to work. Forty years had passed since the little canal around the falls at Richmond had been dug. It was there that the new company began operations, widening and deepening the original channel and moving on up the river. In 1836, fourteen hundred men were employed. The following year the number rose to over three thousand. They were slave laborers working under the supervision of white overseers. They were housed in tents and were well fed. In addition to their wages, which were paid their owners, they earned a small bonus which provided them with enough money to buy whisky off the peddlers who followed the camps. When trouble erupted it was usually traceable to whisky.

Despite good management and the vigorous manner in which the work was being prosecuted, the canal builders ran into unexpected financial difficulties when Virginia, Richmond and Lynchburg defaulted on their subscriptions. The best they could do was to make partial payment. The James River and Kanawha Company had no choice but to agree to such an arrangement. It marked the beginning of years of financial uncertainty, the company never knowing whether the promised and desperately needed payments would be made.

The money dribbled in but seldom on time, and it was only by borrowing funds at 8 percent that the work proceeded. Notwithstanding these handicaps, the 146-mile stretch of canal between Richmond and Lynchburg was opened to navigation in November, 1840.

The first boat through was the *William Henry Harrison*, named for the recently elected ninth president of the United States.[3] Its arrival late in the afternoon of December 3 touched off such a celebration as Lynchburg had never seen. Since morning most of the population had been scanning the canal from the bluffs above the river for a first glimpse of the boat. When it came into view, horses trotting and covered with foam, there was a concerted push to the landing, where city officials, clergy, fire companies, several bands and a company of militia were gathered. Cannon boomed, the militia fired its rifles, bands blared and all Lynchburg cheered. It had been waiting forty-four years for this momentous day.

Several weeks later fast packet service was established between Richmond and Lynchburg, the *John Marshall* and *Joseph C. Cabell* leaving each city three times a week and making the run in thirty hours. They were luxurious boats, the equal of the best seen on the Erie. They were drawn by three-horse teams driven tandem, the driver mounted on the rear horse (the universal canal custom).

The passenger and express business proved to be so profitable that soon three lines were competing for it. The promoters of the Richmond and Allegheny Railroad used that fact as an argument in selling stock in their proposed venture.

The extension to Buchanan was begun shortly and completed in 1851, adding fifty more miles to the length of the canal, almost half of which was achieved by slack-water navigation. The James River and Kanawha Company should have stopped there, but it was determined to reach Covington (Virginia), a prosperous mill town. Obviously it still entertained the hope that it would one day get as far as the Kanawha. In 1858 it bought the North River Navigation, which took it up that stream twenty miles to Lexington.[4] Getting to Lexington cost the company $273,000, very little of which was ever earned back. But the mainline was immensely profitable. In addition to the packets, 195 freight boats were using it in 1854.

Unlike northern canals that had to suspend operations for three or four months when winter set in, the James remained open the year round. In fact, there were very few days when ice froze in the locks so that they could not be used. Floods, however, played havoc with towpath and berm. In 1842 a series of freshets washed out the banks in almost a hundred places. Four hundred men worked through the summer repairing the damages. In 1877, and again the following year, what is

Decaying wreck of the famous packet *John Marshall,* Waterfront Park, Lynchburg, Virginia. *Courtesy National Archives.*

Houseboat on the canal—Kanawha and James rivers.

Confederate troops on their way to the front, James River and Kanawha Canal, Lynchburg, Virginia. *Courtesy National Archives.*

referred to as the "Great Flood" destroyed locks, aqueducts, bridges and channel. In the first years of the Civil War the James and Kanawha Canal was used extensively by the Confederacy for moving troops and supplies, but being located almost at the center of the battleground that Virginia became, it could not escape the ravages of war. General Sheridan and his Union cavalry swept around west of Richmond in March, 1865, and spent a week destroying its weigh-locks, dams and buildings. When Richmond was burned on April 3 of that year, all company books and records were lost.

Prior to the war, in 1859, a railroad from the capital to Lynchburg had been completed. To meet this competition the canal company reduced its tolls drastically and the packet boat companies cut their fares by 40 percent. Thus you could make the forty-four-hour journey from Richmond to Lexington, meals and lodging included, for $5.25. Or, you could go by rail to Lynchburg, catch a packet there and be in Lexington in twenty-four hours—but at twice the cost. Not many people were in that much of a hurry.

From the opening of the Civil War to the surrender at Appomattox railroads were a prime target for destruction. The Lynchburg line had bridges burned and tracks torn up, but after a few weeks it was always put back in commission. Both sides soon became so proficient in destroying the other's roads and in quickly restoring their own that it became almost an art. If very few locomotives and very little captured rolling stock was run off to be used by the enemy it was because northern railroads operated on standard-gauge track—four feet, eight inches— while all southern railroads employed a five-foot gauge.

At war's end the James River and Kanawha Company was still solvent and able to pay the interest on its debt. When the forty western counties of Virginia refused to secede from the Union, the new state of West Virginia came into existence, and with it the confiscation of all rights and improvements the James River and Kanawha Company owned west of the new boundary line. The company weathered the panic of 1873 and the following year began constructing a railroad from Buchanan to Clifton Forge to connect with a new trunkline railroad that was heading for the unexploited coal fields of West Virginia. It appealed to the state and the Congress for financial assistance. Although it received none, it went ahead. It was only 22½ miles from Buchanan to Clifton Forge but they were rugged, costly miles.

The work was well under way when the Great Flood of November, 1877, washed out everything. The canal was so badly damaged that it was at first thought to be beyond repair. After $200,000 had been spent on restoring it, the Richmond and Allegheny Railroad bought the canal and other assets of the James River and Kanawha Company and the latter ended its unhappy existence.

The packet *William Henry Harrison* had been the first boat to reach Lynchburg in 1840. It may have been the last when it took a party of railroad executives up the canal in 1878. No figures are available on the longevity of packet boats but the *John Marshall* was the last survivor of the James River boats. For years it lay rotting at the river's edge in Lynchburg's Riverside Park, regarded with reverence for its historic past. It was honored particularly for that day in May, 1863, when, draped with crepe, a military honor guard standing at attention on deck, it had borne the body of General Stonewall Jackson home from Lynchburg to Lexington.

James River at Richmond after the Civil War.

Although the story of the *John Marshall*'s arrival at Lexington has often been told, southern hearts never tire of hearing it. The young, solemn-faced cadet corps of the Virginia Military Institute was waiting at the landing to receive the remains of their late mathematics professor. They carried the casket to the general's old lecture hall and placed it before the chair from which he had so often addressed them.[5]

Because of the personal nature of their grief, they may not have realized that in the death of Jackson the Confederacy had received an irreparable blow. But young and old alike sensed it when Lee was defeated at Gettysburg two months later. If only Jackson had been there!

8

The Delaware
and
Hudson Canal

THE CANAL MANIA reached its peak in 1830, with no fewer than forty waterways either in operation or under construction. They were built primarily to provide cheap transportation to market of farm produce, timber, coal, iron ore and assorted articles of commerce. They quickly proved that freight could be moved by canal for approximately one-third the cost by wagon. Passenger traffic, regarded at first as a secondary matter, soon became a major source of revenue for the canal companies. That they opened communication with hitherto isolated regions and brought them closer to centers of population was a dividend that could not easily be measured in dollars and cents.

Without exception canal companies suffered from the handicap of discovering that actual construction always amounted to several times the estimated cost. This having become a demonstrated fact, one can only wonder why canal enterprises continued to be launched on such unreliable figures. Perhaps promoters regarded such chicanery as conducive to selling stock, hoping that when the showdown came, additional money could be found somewhere—most likely from the state.

It is even more remarkable that Americans continued to invest millions of dollars in inland waterways construction without giving any thought to the competition the canals might soon have to face from the railroads. In America, Peter Cooper had "improved" his original locomotive and was delighting the directors of the Baltimore and Ohio Railroad with its performance. In England, at Newcastle-on-Tyne, George Stephenson, developing the theories first advanced by his predecessors Thomas Newcomen and James Watt, had invented the world's first successful reciprocating steam engine with tubular boilers, in principal but little different from what it is today. Stephenson demonstrated that, mounted on wheels, it could propel a train of cars running on rails at a modest speed of ten miles an hour.

Stephenson's claims were put to the test when the Killingworth collieries engaged him to construct a railroad from its mine pits to the docks at Newcastle. The experiment was so successful that the great Hetton collieries asked him to build an eight-mile-long railroad for them. These were strictly freight carriers but in 1825 the short Stockton and Darlington Railway, transporting freight and passengers, came into successful operation. It aroused enthusiasm for connecting the manufacturing cities of Manchester and Liverpool by rail.

These portentous forerunners of the future were not unknown in the United States. But they were disregarded. Short railroads might be of some use in comparatively level England, it was said. In this country, with its great distances, mountains and decided pitch to the sea from the Great Lakes to the Carolinas, they could never compete with canals.

And yet, ironically, as is so often the case, it was an American canal company, the Delaware and Hudson, soon to become one of the richest and most prosperous in the country, that dispatched Horatio Allen, a young engineer, to England in 1828 to arrange for the purchase of four of Stephenson's locomotives.[1]

For reasons that company records do not explain, Allen bought only one of

Offices of Delaware and Hudson Canal Company, Albany, 1850.

Stephenson's engines (perhaps it was the only one ready for shipment) and three from the rival concern of Fuller, Rostruck and Company of Stourbridge. One of the latter, named the *Stourbridge Lion*, was to achieve a permanent niche in history as the first locomotive to run on American tracks.

The rags-to-riches story of the Delaware and Hudson Canal, the great anthracite carrier, begins many years before then. Only with wisdom abetted by good luck could it have attracted so many men of outstanding ability to serve it. Horatio Allen was only one. To John A. Roebling was given the task of building the cable-hung aqueduct over the Delaware River, three hundred yards below the mouth of the Lackawaxen River. Roebling's finest achievement was building the Brooklyn Bridge, but the Delaware River aqueduct was proper training for it. It is a monument to his engineering skill. Today, a century and a quarter after it was built, it serves as a highway bridge.

The ubiquitous Benjamin Wright was installed as chief engineer after he finally separated himself from the Erie Canal. Another notable who joined the engineering staff of the D. & H. was John Langdon Sullivan, the fighting head of the Middlesex Canal in its last years. There were others, John B. Jervis for one, who had been Benjamin Wright's chief assistant on the Erie.[2] And, of course, there was wealthy Philip Hone, several times mayor of New York City, and first president of the D. & H.

As late as the early 1800s mountainous, thinly populated northeastern Pennsylvania, with its timber and rushing streams, was still regarded by outsiders as a semi-wilderness. Game abounded and there was little communication with the

Philip Hone, several times mayor of New York City and president of the Delaware and Hudson Canal Company. *Courtesy Museum of the City of New York.*

John Roebling's suspension aqueduct across the Delaware River, built for the Delaware and Hudson Canal.

outside world. Its remoteness and primitive life appealed to Maurice Wurts, a Philadelphia merchant. Every autumn he journeyed north to the Lackawanna River to spend a month or two fishing and hunting with Dave Nobles, his friend and backwoods guide, while his brother William kept store in his absence.

In the course of three or four years, roaming over wilderness trails, sleeping out wherever night found him, there wasn't much of Wayne and Susquehanna counties that Maurice Wurts did not see. On arriving on the Lackawanna in 1813 he found Nobles facing imprisonment for debt and about to lose his property. In what must be regarded solely as a friendly gesture, without thought of gain, he took title to Nobles' land and relieved him of his debts.

From the many outcroppings of so-called stone coal it had been assumed for years that fields of anthracite existed beneath the surface. But no one was interested; as fuel, stone coal had no value, nor was it likely ever to have with wood so cheap and abundant. It was widely believed that it would not burn at all, though experiments in Philadelphia that winter proved it was combustible if given proper draft. Why there should have been any question about it is a mystery, for during the Revolution small quantities of anthracite had been floated down the Susquehanna River to the Carlyle Arsenal for use in the manufacture of arms.

Maurice Wurts must have become apprised of this, for during 1814 he took Dave Nobles into his employ and together "they explored a great expanse of the Lackawanna Valley, mapping coal outcroppings and purchasing what land they thought of value at from 50¢ to $3.00 per acre."

The Delaware and Hudson Canal: A History, published by the Wayne County Historical Society, leaves no doubt that Maurice Wurts was by now committed to embarking in the coal-mining business. Mining a few tons of anthracite was a simple matter, but transporting it to Philadelphia, where it could be exhibited, was a problem.

> In the spring of 1816 he attempted to float a raft-load of this coal down Jones Creek, a tributary of the Wallenpaupack, but fortune was not with [him] for the raft struck some rocks and was quickly broken up. Although the accounts are somewhat vague and contradictory, [he] seems to have succeeded in hauling a small quantity over the old Wyoming Road the following year, rafting it down the Wallenpaupack to the falls where it was unloaded, hauled to the Lackawaxen near Paupack Eddy and again loaded on a raft for the long voyage to Philadelphia.[3]

The Wurts brothers, Maurice, William and John, the last a lawyer, were disappointed in the lack of interest their exhibition aroused in Philadelphia. It was pointed out to them that anthracite, whatever its merits as fuel, could have no future unless some cheap and adequate means of transporting it from the mines to the city was discovered.

It was only one of the many rebuffs Maurice Wurts was to encounter. But he was not easily discouraged. Having come to believe that cheap transportation held the key to the great enterprise he envisioned, he was determined to find a way. In exploring the valley of the Delaware many times, he had become aware of the broad trough that once connected it with the Hudson River. Hopefully, he turned to it, making a rough survey as he trudged north up the Neversink River. After crossing the Summit, he followed Rondout Creek all the way down to the Hudson and tidewater below Kingston. He was a hundred and some miles from where he began his tour of the valley, and the course he had taken was one that a canal might take. It would open the way to New York City, and Lackawanna coal would have that market to itself. From the mouth of Rondout Creek, a steamboat could tow a string of canal boats downriver to the wharves of Manhattan.

Before that dream could be realized many serious problems would have to be overcome. The fast-flowing Delaware River would have to be bridged by a costly aqueduct at or near the mouth of the Lackawaxen and a way devised to reach the coal fields which lay beyond the intervening Moosic Mountains, a thousand-foot-high barrier, about where the town of Carbondale now stands.

Not being an engineer himself, Wurts turned to Benjamin Wright, then nearing the height of his fame as chief engineer of the Erie Canal. Wright sent him two of his young associates, John B. Mills and Edward Sullivan, to make a detailed survey for a canal from the Delaware to the Hudson. Their findings were so favorable that Wright himself came down from Syracuse and made a personal inspection of the project. He made several minor changes in what had been done and advised the Wurts brothers that for the present the canal terminus should be located at the mouth of Dyberry Creek where it flows into the Lackawaxen. (This was the site of the future Honesdale.)

Several months later Wurts received a set of plans from Wright for a series of inclined planes, powered on some levels by stationary engines and on others by hydraulic pressure, by which dump cars could be loaded at the mines and lowered and raised up and down the mountain slope to the canal head. When James A. Sullivan joined the engineering staff some time later, he insisted dogmatically that Wright's system of inclined planes would not work. Instead, he contended that by lengthening the canal several miles and digging a short tunnel, the coal fields could be reached without difficulty. This was absurd and must be attributed to his jealousy of Wright. Fortunately Wright's ideas prevailed.

The Wurts brothers were not wealthy men, but believing they had a fortune within their grasp, they were willing to risk their all. John Wurts had gotten into politics and was attracting some attention. From the state of Pennsylvania he obtained for his brother Maurice and his heirs and assignees authorization to improve the navigation of the Lackawaxen River. Without opposition this act was approved by the General Assembly on March 13, 1823. On April 23, a little more than a month later, he secured from the state of New York a charter for the Delaware and Hudson Canal Company to construct a canal "from Rondout [Kingston] on the Hudson River to Saw Mill Rift [the state line] on the Delaware River." The route it specified was "up Rondout Creek through the valley to the west of the Shawangunk Mountains, thence down the valley of the Neversink River to the Delaware."

The Lackawaxen Coal Mine and Navigation Company was incorporated by John Wurts. Its title was impressive but it was strictly a family-owned corporation without a dollar in its treasury. However, it was something that could be sold to the Delaware and Hudson Canal Company at a handsome profit, once the latter was organized and financed.

In piecing together the story of the Wurts brothers, Maurice appears to have been the one who led the way. Besides energy and integrity, he possessed an invaluable sense of showmanship. Late in 1824 he rafted a few tons of anthracite down to Philadelphia, where it was transferred to the sloop *Toleration* and forwarded to New York City. Establishing his headquarters there in Sykes' Hotel, he published Benjamin Wright's report and advertised that the charter granted the Delaware and Hudson Canal Company by the state of New York was on display at the bar of the Tontine Coffee House and that copies were available to interested parties and could be had by "calling upon a committee of the Lackawaxen Company at Sykes' Hotel."

A brochure Wurts issued, extolling the financial rewards to be gathered from building a canal connecting the Delaware with the Hudson, aroused the interest of Philip Hone, the wealthy mayor of New York. He was present at what Maurice Wurts described as a "coal-burning exhibit" at the Tontine Coffee House on December 18, 1824. Hone was so impressed by what he saw that he announced his unqualified support of the D. & H. The rest was comparatively easy; Hone had been so uniformly successful that where he ventured, other men of means were ready to follow.

Subscription books for the purchase of stock in the Delaware and Hudson

Canal Company were opened on January 7, 1825, and by early afternoon the entire issue had been sold. The corporation was formally organized on March 8, Philip Hone being elected its first president. Although he resigned his office a few months later, feeling that his duties as mayor of New York demanded all his time, he remained for years on the canal company's board of managers.

Benjamin Wright was made chief engineer. With him he brought to his new post John B. Jervis, for whom Port Jervis was named, another Erie engineer, who was to succeed Wright two years later.

In June the two original Wurts companies were merged, the brothers being paid $40,000 in D. & H. stock for the rights and privileges of the Lackawaxen Coal Mine and Navigation Company.

Work of constructing the canal was about to begin. At noon on July 13, 1825, a great crowd was gathered on the summit between what today are the towns of Ellenville and Wurtsboro to partake in the opening ceremonies. Mostly they were farmers from Sullivan, Orange and Ulster counties who had invested in the canal. President Hone and his party from New York were on hand. They had spent the previous evening in Ellenville, little dreaming that the rustic village and surrounding valleys would one day be a great vacation land for New York's huddled masses.

There were prayers, singing and speeches both before and after President Hone turned the first spadeful of earth. In the course of his remarks he mentioned that the D. & H. was the first canal to wind through New York that was not state-owned. It remained the only privately owned New York State waterway.

As usual, the labor contractors brought in hordes of immigrant Irish and several hundred Germans as well. The Irish fought the Germans and fought among themselves, pitched battles in which dozens of men were maimed. "Of course the brawling slows up the work," Jervis complained to a reporter for the *Kingston Advocate*. "No canal was ever dug through pleasanter country—no swamps or muck to contend with—no extremes of weather. But that doesn't mean anything to these club-swinging Irishers. I don't know what they've got to fight about. They don't need a reason; they fight just for the hell of fighting."

The German laborers and the "wild Irish" could not be housed together, but although placed in separate barracks several miles apart, few weekends passed without the Corkonians storming forth to battle. They became the terror of the countryside, raiding orchards and gardens, taking whatever they wanted and destroying much of what they couldn't carry off. There was no way to control them, and no sheriff was rash enough to try. But as workers their equal was not to be found. How, in the days before the invention of construction machinery when the only power was muscle power, American canals could have been built without them, historians have refused to speculate.

Notwithstanding its labor problems, work on the canal progressed so rapidly that sixteen months after the day the symbolic first spadeful of earth was turned, the section between the Hudson and the Delaware, sixty miles in length, was completed. But company funds were exhausted. Worse still, the easygoing part was over and it was forty-eight tough and costly miles to Dyberry Creek (Honesdale), the announced destination. In extremity the Delaware and Hudson Canal Company sought aid from the state, and the legislature granted the company a loan of $500,000 and permitted it to borrow an additional $300,000 from private sources.

Cable construction of Roebling's Delaware River viaduct. *From Thompson,* A Short History of American Railways.

With its money problem solved, at least for the time being, work was resumed in March, 1827; contracts were let for the Delaware section and the Lackawaxen section as far as the Narrows. The following month contracts for thirteen additional miles were signed, bringing the work to within seven miles of the forks of Dyberry Creek. In a report to its stockholders the company explained that: "It is determined, after much reflection and examination, to stop the canal at Dyberry Forks and from thence to construct a railroad [Benjamin Wright's inclined gravity] to the coal mines, a distance of fifteen miles."

Along the rocky Delaware and Lackawaxen numerous cliffs rise abruptly out of the water. To create a shelf at the river edge on which a canal could be built, D. & H. contractors set their powdermen to work blasting off the crest of the cliffs and causing the falling rocks and debris to pile up below. They were so successful that they built a passageway forty feet above the white water of the rushing streams. It was hazardous work. The uncertainties of black-powder blasting—fuses sputtering out and frequent delayed discharges—took the lives of many men.

It also provided the wild Irish with some amusement. They delighted in raining down rocks on passing raftsmen and lumbermen who had been using the streams for years and felt they had a prior right to them. The rivermen howled that the blasting was deliberately timed to catch a passing raft. The feud got to the courts when the canal company, to avoid an unnavigable turn in the river a mile above the Narrows, built a dam across the curve and dug a new channel. The raftsmen claimed that the dam was improperly constructed and made rafting unnecessarily dangerous. The D. & H. settled some small claims against it and the matter was dropped.

In October, 1828, water was let into the canal and the company announced that the 108 miles from the Hudson to Honesdale were open. This, of course, did not include the great Delaware Aqueduct which was not to become operable for another twenty years. While little progress had been made in building the "Gravity," the following figures reveal how vast was the amount of work accomplished on the canal proper:

107 LOCKS
22 AQUEDUCTS
110 WASTE WEIRS
16 FEEDER DAMS
22 RESERVOIRS
136 BRIDGES

The channel was originally dug only four feet deep, twenty feet wide at the bottom and thirty-two feet at the waterline. Over the years it was widened and deepened on three different occasions, and in its final state it was navigable for boats of a hundred tons capacity.

The first boat to navigate the entire length of the canal was the packet *Orange.* It left Rondout on October 16 with many company officials on board, and two days later reached Honesdale, 972 feet above sea level, and found it a thriving village of several hundred people, where a short time before there had been only wilderness.

The *Orange* was followed by a flotilla of small freight boats. Into them went the tons of coal that had been hauled over the unfinished road from the mines by sled and wagon. As soon as they were loaded, the boats began the long return voyage to the Hudson. At every lock they were cheered by groups of rustics who had gathered to see them pass. For the last three miles the canal used Rondout Creek for its channel. Nearby Kingston had anticipated the arrival of the boats, and when the first, the snub-nosed *Superior,* came into view, the great crowd that had gathered at Rondout greeted it with prolonged cheers, the Kingston militia saluted it with several volleys of musketry and the Kingston band blared forth martial music.[4] There was much fanciful oratory, but even the most extravagant predictions of what the future held for the Delaware and Hudson Canal and the surrounding country were to be surpassed.

The sloop *Toleration* (which chanced to be the vessel that Maurice Wurts had used to ship his sample of Lackawaxen coal from Philadelphia to New York) was waiting; the canal boats transferred their cargo to her and she set off down the Hudson. On December 10, five days later, the first shipment of D. & H. anthracite reached New York.

THE STOURBRIDGE LION
First English locomotive on Delaware & Hudson track

(*From a drawing.*)

The *Stourbridge Lion. From* Leslie's Weekly, *August 8, 1829.*

Philip Hone proved that he knew something about public relations by shipping fifty tons of what the press had dubbed "black diamonds" up the Hudson to Albany "to insure the comfort of Governor-elect Martin Van Buren and the legislature." The gesture bore fruit, for the D. & H. received a second loan of $300,000 from the state, which may explain the company's decision to send young Horatio Allen to England on his locomotive-buying expedition. As previously mentioned, he purchased four of the "primeval monsters" and they arrived in New York the following spring. Three of them were stored in an East Side warehouse. What became of them is unknown. But the fourth, the *Stourbridge Lion* (so named because of the lion's head painted on the circular front end of the boiler), was taken up the Hudson as soon as the ice was out of the river and was unloaded at Honesdale early in August.

Three miles of track had been made ready. But the *Lion* weighed eight tons, almost twice its previously estimated weight. Fears were expressed at once that the track would not be able to sustain it. The rails were only strap iron, laid on hemlock stringers. Horatio Allen was on hand and volunteered to take the engine on its trial run. Of that experience he later wrote:

I had never run a locomotive nor any other engine before. But on August 9th, 1829, I ran that locomotive three miles and back to the place of starting, and being without experience and without a brakeman, I stopped the locomotive on its return at the place of starting.

The line of road was straight for about 600 feet, being parallel with the canal, then across Lackawaxen Creek on trestle-work about 30 ft. above the creek, and from the curve extending in a line nearly straight into the woods of Pennsylvania.

When the steam was of right pressure . . . I took my position on the platform of the locomotive alone, and with my hand on the throttle-valve, said "If there is any danger in this ride, it is not necessary that more than one should be subjected to it."

The locomotive having no train behind it answered at once to the movement of the valve; soon the straight line was run over, the curve (and trestle) was reached and passed before there was time to think . . . and soon I was out of sight in the three miles' ride alone in the woods of Pennsylvania.[5]

But watchers had seen the trestle sway and heard it creak and groan. Jervis examined the track and found that some of the rails had been knocked out of alignment. At his direction the *Stourbridge Lion* was dragged into a vacant shed and left to rust away. It never ran again. Later its boiler and other parts were sold. But after the passage of many years it was reassembled by the D. & H. and given to the Smithsonian Institution.

For the first two years the canal was in operation the larger part of the D. & H.'s earnings came from hauling wood rather than coal. But with the completion of the Gravity, that changed at once. Dividends rose to 11 percent and never dropped below that figure even though the company spent thousands of dollars in opening navigation to the Susquehanna by way of the Lackawaxen, thereby capturing the business of the Pennsylvania Coal Company, the biggest operator in the Susquehanna fields. That profitable arrangement lasted until the Erie Railroad bought the Pennsylvania Coal Company. But by then the D. & H. had more of its own business than it could handle. Dividends on its capital stock seldom fell below 20 percent, and in 1872, the peak year, when it carried 2,930,333 tons of coal, they reached 31 percent.

Numerous boatyards came into existence along the canal, most of them financed by the company, which was their principal customer. All told, the D. & H. bought something like nine hundred boats and then sold them on easy terms to men who were anxious to engage in boating on the canal. The life of a canal freighter was estimated to be twenty years. The average cost was $1,500. On the terms offered by the company a man could pay for his boat in ten years. After that what it earned was all profit to the boater.

Great deposits of trass, so necessary for making hydraulic cement, had been found near Rosendale when the canal was being built. Since then a prosperous cement industry had sprung up there. In line with company policy to advance the interest of the country through which it passed, the D. & H. was happy to support it. In deference to public opinion, it did not operate on Sunday. At first its locks were numbered in sequence westward from Eddyville (Rondout) but they soon acquired local names, usually of some farmer or villager. The company adopted these and printed them on its table of distances. The country was too thinly settled for its passenger business to be profitable, but the D. & H. maintained it as a convenience. For these and many other reasons the Delaware and Hudson won a measure of cooperation, even friendship, such as no other canal ever enjoyed.

For twenty-seven years—from 1831 to 1858—John Wurts served as company president. His brother Maurice, the pioneer, had died in 1854. By then it was obvious that the railroads were putting the canals out of business. The managers of the Delaware and Hudson Canal Company avowed that they would never turn to steam, but in January, 1899, they applied for and received permission to drop "Canal" from the corporate title. As the Delaware and Hudson Company they bought the little Albany and Susquehanna Railroad, and then proceeded to buy other roads, most of them northward in New York State until they finally had trackage all the way to Canada.[6] The doom of the old canal, which had outlived so many of its contemporaries, had sounded. On June 13, 1899, the water was let out and it was abandoned.

9

The Lehigh Valley Canal

WHERE THE DELAWARE RIVER breaks out of New York State and begins its long journey to tidewater below Philadelphia, it becomes, with all its twistings and turnings, the boundary that separates New Jersey from Pennsylvania. In the heyday of the canals no other river played so important a part in their navigation as the Delaware. Canals using the Delaware included the Schuylkill Navigation, the Delaware and Hudson, the Delaware and Raritan, the Lehigh and the Morris.

In previous chapters the story of the Schuylkill and the Delaware and Hudson has been covered. Of the three remaining canals that used the Delaware, or at least crossed it, the Morris was chartered in 1824, the Lehigh a year later (although the Lehigh was in operation long before the Morris was dug) and the Delaware and Raritan came into being some years later.

If a canal had character it was invariably because it reflected the personality of the hard-headed visionary and fanatic who conceived it—Maurice Wurts, for instance. The Lehigh Canal produced his equal in the person of Josiah White. He was of Quaker stock and possessed little formal education, but was shrewd, determined and not easily talked out of what he believed in.

The first record we have of him as a young man is as the proprietor of a small shop in Philadelphia that manufactured wire nails, a new invention recently on the market which was having a difficult time replacing the hand-forged nails that

man had been using for centuries. White wasn't prosperous and it didn't appear that he ever would be if he remained in the nail business. In 1817 he closed up shop and set out to discover if an idea that had been nagging him ever since he had witnessed Maurice Wurts' "coal-burning exhibition" was worth pursuing.

Convinced that "stone coal" possessed many advantages over wood and bituminous for heating purposes, he was equally confident that it would soon find a wide market. But first some inexpensive means of transporting it to Philadelphia would have to be found.

Back in 1792, before White was born, Jacob Weiss, Charles Gist and several others had purchased ten thousand acres of coal-bearing land in the upper Lehigh Valley at Mauch Chunk. They had patented the tract and organized the Lehigh Valley Coal Mine Company. Nothing was done about developing the property, the owners apparently having concluded that they had made a bad investment and that there was no point in mining coal for which no means of transportation could be found. But though long moribund, the company had continued to pay its taxes—which couldn't have amounted to much—and was still in existence.

Josiah White had heard the story. It occurred to him, poor as he was, that the property might be purchased for little or nothing. Before contacting the owners, he decided to make a personal examination of the company's holdings. His friend Erskine Hazard agreed to accompany him.

Hazard got the loan of a set of levels that had been used on the Union Canal; White borrowed a horse, bought a few supplies and a rifle. Their trip up the Delaware to Easton was uneventful, but there they got their first look at the Lehigh River. Even at its confluence with the Delaware it was a wild mountain stream.

"A few miles above Easton," Josiah White recalled, "the Lehigh was pocked with white water at almost every turning. To navigate it seemed impossible. But I learned that in seasons of high water a few arks and rafts made it down to Easton."

He and Hazard followed the river to Mauch Chunk, forty-six miles or more. After exploring the property they had come to see they went on to White Haven, another twenty-five miles, and from there to Stoddartsville, thirteen miles further, before they turned back.

"As I recollect it, we saw only one house between White Haven and Stoddartsville," White recalled. "It rained most of the time. We slept on the ground, and it was wet. We had eaten ourselves out of grub and were living on the game we killed."

At the time there couldn't have been more than three hundred people living in what today is Carbon County, with a flouring mill and a gristmill at Mauch Chunk and another at White Haven. Although it was less than a hundred miles from many centers of population, it was isolated, primitive, backwoods country, with few roads, most of them no more than trails. But Josiah White and Erskine Hazard were not concerned about the backwardness of the region; they had examined and measured the land they had come to see and found that it exceeded their expectations. Better still, White was confident that with some improvements he could tame the Lehigh and make it usable for at least downriver navigation.

That two young men, bereft of worldly goods, dared to hope to gain control of ten thousand acres of rich coal lands seems incredible. But there were some factors in their favor. Half of the men who had launched the Lehigh Valley Coal

The Lehigh Valley Canal.
Courtesy Edwin P. Alexander Collection.

Lehigh Canal—Easton to
Bethlehem.

Mine Company had died; those who remained were so heartily sick of baby-sitting
with an investment that had been disastrous from the start that they were ready
to listen to anyone who had a proposition to offer. So they listened to Josiah White
and his partner. Out of those conferences came the famous "ear of corn" agreement, the like of which had never been seen before (or since). For a rental of
one ear of corn annually White and his partner Hazard were given a twenty-year
lease on the coal lands, with the stipulation that they were to pay the taxes on
the property, and after three years of occupancy *begin* shipping forty thousand
bushels of coal to Philadelphia.[1]

Because White and Hazard had no money and an acquaintance named Hauto
had a little, they took him in as a partner, and the three hurried off to Harrisburg
to obtain from the legislature the exclusive privilege of improving the Lehigh.
Successful, they organized and incorporated the Lehigh Coal Company. Miraculously they managed to dispose of $50,000 worth of stock. In the spring of 1818,
with only a dozen laborers at first, Josiah White went to work to bend the Lehigh
River to his will. The number of men employed rose rapidly. By the end of the
year five hundred or more were toiling in the river. The wage was seventy-five
cents a day and anyone who wanted a job could find one.

White drove them hard, but no harder than he drove himself. "I was in the
water with them most of the time," he reminisced years later. "I didn't know
anything about blasting nor did they. But we were lucky; as I recollect we lost
only two men."

In saying that he knew nothing about blasting he was unduly modest. He may have been inexperienced in the beginning but he soon became so proficient in making wax-paper fuses that they seldom failed to "blow" or fire when ignited. And that was the secret of successful blasting.

White has been called a fanatic. Certainly he had an inventive mind. The idea of making downriver navigation of the Lehigh possible by creating a series of artificial freshets was original with him.

Although few believed that his idea would work, he went ahead with it and at a mean stretch of rapids he built a V-shaped dam across the river and equipped it with gates at the point of the V. When the dam was filled, he opened the gates. Temporarily his "flood" raised the level of the river several feet and an ark passed over the rapids without hindrance. Thus was demonstrated the practicability of what came to be called the "bear-trap gates," for which White is justly famous.[2]

White was able to forward a thousand bushels of Lehigh coal to Philadelphia in December, 1819. This token shipment was for publicity purposes rather than profit. It enabled White and his partners to dispose of more stock in their company.

Pool navigation of the Lehigh was completed the following year. Josiah White, still experimenting, found that two, three, even four arks could be hitched together and sent downriver at the same time, a single crew managing the little flotilla. In addition to the obvious economy, there was a further saving in having fewer men to haul back to Mauch Chunk from Philadelphia by wagon.[3]

Completion of a tramway from the mines to the river at Mauch Chunk expedited loading the arks. A vast amount of anthracite, timber, barrels of flour and other farm produce was shipped down the Lehigh that year. Great as the tonnage was, it doubled by 1823. Upwards of a thousand tons of coal reached Philadelphia and glutted the market. The retail price fell from $5.50 to $3.00 a ton.

Knowing that the output of the mines could be doubled, even tripled, White and his partners began casting about for an additional market. Providentially it seemed, the Morris Canal Corporation was organized in 1824 to construct a waterway between Easton and New York harbor. This meant that Mauch Chunk coal could be shipped from the mines to New York City simply by transferring it from the arks to canal boats at Easton. The empty arks could continue on down the Delaware and be sold for their lumber, as heretofore. A great market would be opened if New York City could be induced to use coal.

In 1825 the Lehigh Coal and Navigation Company (the change in its corporate title had occurred in 1820)—now owners of the Mauch Chunk mines— announced that it was going to build a canal up the Lehigh River from Easton to Mauch Chunk and White Haven, and later extend it to Stoddartsville, a total distance of eighty-five miles or thereabouts. It was a major undertaking estimated to cost well over a million dollars. Many friends and well-wishers of White and his partners shook their heads and said, "It's the first incautious step the boys have taken." Other skeptics predicted that Josiah White and his partners would lose their shirts.

When work began at Mauch Chunk in April, 1825, White was head of construction, as might have been expected. Again, he could say he was inexperienced and knew nothing about building a canal. But he was of the breed that learns quickly.

Like the James and other canals, the Lehigh was a combination of canal and slack-water pools. When finished, there were forty-six miles of canal and ten miles of slack water between Mauch Chunk and Easton.

The canal was considered complete in late June, 1829. The water was let in on July 1 and reached Easton in midafternoon. Josiah White was on hand, as were his partners, a group of notables and the ubiquitous Canvass White who had been installed as chief engineer in 1827 and completed the Lehigh.

The packet boat *Swan* had been brought up the Delaware from the Schuylkill for the occasion. She was let into the canal. The assembled guests boarded her, and when they were seated at points of vantage, the captain signaled the hoggee on the towpath. The mules leaned against their collars and the first voyage up the new waterway had begun.

It could not have been a day of unalloyed joy for Josiah White. Gazing across the Delaware at the village of Phillipsburg, New Jersey, he looked for what he knew was not yet there and would not be for another two years: the Morris Canal on which many of his hopes were based.

As canals went, the Lehigh was not long. But its dimensions were large. White had built it for the future, for a day when boats much larger than any then in commission would be using it. It was sixty feet wide at the water level, forty feet wide at bottom and five feet deep. Its stone locks were one hundred feet long and twenty-two feet wide—large enough to accommodate boats carrying 100 or 150 tons.

Before the 1829 season was over, a line of packet boats was plying the Lehigh from Easton to Mauch Chunk, where another line provided thrice-weekly service to White Haven. The canal appeared to be what the country had been waiting for. It began to boom, new villages sprang into existence and the price of good farm land doubled. The arks that had been the only means of transportation in the days of the bear-trap gates began to disappear. The few that remained avoided the canal to escape the tolls.

It is impossible to mention the valley of the Lehigh without there popping into mind almost automatically the name of the great coal carrier, the Lehigh Valley Railroad. But the railroad had no connection with the Lehigh Coal and Navigation Company, owners of the canal. When the railroad put its tracks into Carbon County and spread out its branch lines, it came into fierce competition with the rival company. But although the Lehigh Coal and Navigation Company was seriously affected by the conflict, the Lehigh Valley Railroad never was able to put it out of business. In fact, after the coming of the railroad the canal enjoyed some of its most prosperous years, due in part to the opening of the state-owned Delaware Division Canal, which permitted it to reach many New Jersey industrial towns.

A bad flood in 1841 put a large section of the canal below Mauch Chunk out of business for that year. The White Haven division was so seriously crippled that service was not restored for three years. Violent storms often tore down the Lehigh Valley, leaving destruction behind them. The great flood of 1841 had been memorable but it wasn't much compared to the one that devastated the valley in 1862 and wrecked the White Haven division so completely that it was permanently abandoned; between Mauch Chunk and Easton twenty miles of the canal were ripped asunder. Old-timers called it the most terrible flood that ever struck the Lehigh.

The railroads also were severely damaged; roadbeds were washed out and bridges carried away. For several months during 1862 no trains ran. In his old age

Barges, lower left, on the Lehigh Valley Canal, Mauch Chunk, Pa., c. 1870. *Courtesy Edwin P. Alexander Collection.*

Barge going downstream, probably
Bucks County. *Courtesy Edwin P.
Alexander Collection.*

Steam tow of five barges, Delaware and Raritan Canal. *Courtesy Edwin P. Alexander Collection.*

Levi Gauss, who lived on the opposite side of the river below Mauch Chunk, liked to recall the third day of the flood, its worst:

> Along the front at Mauch Chunk you couldn't see nothing but the coal chutes; everything else was under water. Loaded boats had been swamped and sunk to the bottom. Folks up toward White Haven had left their homes for higher ground before their cabins floated away. Seemed like the flood was tearing everything loose. Uprooted trees, smashed cabins, drowned cattle, chicken coops, outhouses—everything was being swept down river. There ain't been nothing like it since.[3]

Despite the calamities from which it suffered, the Lehigh Canal was one of the most profitable dug in America. In the decade before it came into competition with the railroad it never failed to carry less than three-quarters of a million tons. In 1860, its peak year, the total reached 1,338,875 tons.

The Lehigh has had a long life. Well into the twentieth century, mules were still plodding along the towpath between Siegfried's Bridge and Easton.

From the beginning anthracite was screened and marketed in two sizes; the larger "egg" coal and the smaller "chestnut" (the latter for home-heating). "Buck-wheat" and "slack" coal—the latter little more than dust—had been pushed aside as valueless and left to be washed by the rains into the river. With the introduction of forced-draft furnaces in big industrial plants it became salable as low-grade coal. The Lehigh was clogged with what had always been regarded as "waste." Recovering it by dredging quickly became a major operation. It was so successful that in 1924 upwards of 175,000 tons of waste coal were shipped down the canal.

No one would have been more amazed than young, ambitious Josiah White, who had first set foot in the valley of the Lehigh more than a century ago, when few people really believed that anthracite would burn.

Repairing channel damaged by freshet, on the Lehigh Canal.

10

The Morris and Delaware Division Canals

ONLY A BACKWARD glance is required to reveal how unequal was the struggle for supremacy between the canal and the railroad. After 1835 there was little canal companies could do to improve their service. While larger boats could be built, their size had to be limited to the dimensions of the channel on which they operated. Speed could not be increased, since a boat moving at faster than the standard four miles an hour would create a backwash that would tear holes in berm and towpath. Nor could much be done to make passenger-carrying packets more comfortable. Railroads, on the other hand, were constantly being improved.

In 1830 there were only twenty-three miles of track in the United States; ten years later the figure had risen to 2,818 miles.[1] Stagecoach and canal interests banded together to excoriate the steam trains and their owners. "Take warning, America," screamed one of the pamphlets they scattered broadside. "The steam locomotive is the Agent of Satan. It will maim and destroy you and your children." The clergy joined them, protesting from a thousand pulpits that "Operating railroad trains on the Sabbath is a desecration of the Lord's Day." The protests against running trains on Sunday became organized, with the formation of Sabbath Day Committees and Alliances. The railroads were forced to take heed; some roads canceled their Sunday trains; others reduced the number of them, sometimes an-

Inclined plane on the Morris and Essex Canal, Stanhope, N.J., c. 1910. *Courtesy Edwin P. Alexander Collection.*

◄

The so-called humpback bridge lock at Grantville on the Morris Canal. *Courtesy Little Falls, N.J., Public Library.*

Horatio Allen, in his old age. The first man to drive a locomotive in the United States.

nouncing that "Complying with the Sabbath laws, no train shall leave its terminal
before 6 P.M."

Due to faulty equipment and the carelessness of employees, numerous accidents occurred on the railroads, injuring and killing an unconscionable number
of people. But, though under more or less constant attack, the railroads threatened
the future of the canals. Down in Charleston, South Carolina, Horatio Allen, of
Stourbridge Lion fame, had been appointed superintendent of the Charleston and
Hamburg Railroad, then under construction. A locomotive of his design was built
at the West Point Foundry at Cold Spring, New York. Quaintly named *Best Friend
of Charleston,* it was shipped by steamer to that city. There in December, 1830,
with Allen at the throttle, it pulled the first train of cars ever moved by steam in
the United States.[2]

Proving that it was not necessary to import locomotives from England was a
big plus for the promoters of American railroads. But unfortunately, the *Best
Friend* exploded its boiler a few months later, which temporarily cooled off the
enthusiasm for railroads and the investing public turned again to the canal companies. Preliminary surveys for a canal extending from the mouth of the Lehigh
River to New York harbor had been made. The promoters, feeling that the time

Lock at Green's Bridge, Phillipsburg, N.J., c. 1900. *Courtesy Edwin P. Alexander Collection.*

was ripe for organizing the company and putting its stock on the market, applied to the New Jersey legislature for a charter. The privilege was granted and the Morris Canal and Banking Company came into existence.

Constructing a canal between the Delaware and New York harbor was a sound idea.[3] It would reach the industrial and iron mining region of New Jersey and connect with the coal-carrying Lehigh, insuring a steady volume of tonnage west to east, while the westbound business from New York promised to be substantial.

As the crow flies, the distance between Phillipsburg, across the Delaware from Easton, and Newark Bay was a fraction better than 55 miles; the distance by canal totaled 102 miles. The additional mileage resulted from taking the least expensive way through the hilly, semi-mountainous country around Lake Hopatcong, a 914-foot climb. Going down the opposite side there was a drop of 760 feet; a total rise and fall of 1,164 feet.

As originally built, the Morris Canal took off from Upper Newark Bay, followed the Passaic River to Little Falls, which it crossed by viaduct, and on to Boonton, Dover and the southern tip of Lake Hopatcong, from where it pointed in a southwesterly direction for Phillipsburg and the Delaware River. Some years later, after it had been demonstrated that the Newark Bay terminal was costing the canal company business, the channel was extended twelve miles across Bayonne Neck to Jersey City and New York Bay.

The Morris Canal and Banking Company was incorporated for $2,500,000. A stock issue in that amount was quickly sold—or to be more exact, subscribed for, which was considerably different. So many subscribers defaulted on their pledges that the company had to go ahead with little more than a million dollars at its disposal. How it was to build a canal estimated to cost more than twice that amount must have given the directors grave concern. Attempting to economize, they made the mistake of building a canal that was too small to produce a profit.

Instead of tackling the costly problem of getting the canal over the mountains to Lake Hopatcong, they began construction at Newark Bay late in 1824. The channel was only thirty-one feet wide at top, twenty feet at bottom and four feet deep. The locks they installed were so small they could not handle boats carrying more than twenty-five tons. Obviously they could not accommodate the big boats

Loaded boat ascending a plane on the Morris Canal. *Courtesy Passaic Historical Library.*

Barge or boat in lock on the Morris and Essex Canal, c. 1870. *Courtesy Edwin P. Alexander Collection.*

Inclined plane on No. 11 East, Morris Canal, c. 1900. *Courtesy Edwin P. Alexander Collection.*

Aqueduct over Passaic River at Little Falls, N.J.

Lock 15 above Newark on the Morris Canal, New Jersey. *Courtesy Little Falls, N.J., Public Library.*

being used on the Lehigh. Thus, automatically, any hope of through traffic from that source was ruled out, or any Lehigh traffic at all, until the Morris Canal Company built a fleet of small boats. It is difficult to understand how even that procedure could have been justified, for it necessarily entailed the cost of transshipping Lehigh coal at Easton. Furthermore, a small boat would have to be manned by the same size crew as a large one.

But if the directors of the Morris followed a course that was sometimes incomprehensible, they made no mistake when they engaged Professor James Renwick of Columbia University, an Englishman, to devise a practical means of getting the canal up and down from Lake Hopatcong, a matter of more than sixteen hundred feet in ninety miles.

Since the average lift and fall of a canal lock was six feet, Renwick convinced the director of the company by some simple mathematics that getting through to the Delaware by a system of lockage would require the building of upwards of three hundred locks, the cost of which would run into millions of dollars. Instead, he advised a series of inclined planes similar to those his friend Benjamin Wright had constructed for the Delaware and Hudson Canal. Having no better choice, the directors authorized Renwick to go ahead.

In the simplest terms, the inclined plane was nothing more than a short boat-railway up which a canal boat could be raised or lowered by cable (hempen ropes were used in the beginning), an ascending boat counterbalanced by one that was descending. The power was supplied by a stationary engine, or, as in the Morris Canal, by water power.

Renwick found it necessary to build twenty-three inclined planes, but only a like number of locks. It was one of the finest engineering feats in canal history. On November 1, 1830, long before he was finished, the eastern slope of the canal, between Dover and Newark, was opened for a trial run. Several boats loaded with ore and finished iron left Dover and passed the planes without delay or mishap. On May 20, 1832, the canal was declared officially open. Two boats loaded with Lehigh coal, which they had taken on board at Phillipsburg, made it all the way through to tidewater.

They were small company-owned boats, so-called flickers. The excitement aroused by what was considered to be the successful opening of the canal gave Morris stock a temporary boost. However, it became painfully clear that if the canal was to be operated at a profit it would have to be enlarged as well as extended across Bayonne Neck to New York Bay. But there was no money in the treasury to pay for such improvements. Unquestionably it was the search for funds that led the Morris Canal and Banking Company astray.

The old California saying that "more gold was put into the ground than was ever taken out of it" was applicable to investing in canal stock, many investors losing every dollar they risked. Some canal promotions were outright swindles. But in the majority of cases failure was due to ignorance and honest miscalculation. Behind the Morris Canal and Banking Company were men of good reputation, prominent in their communities, but in a mad scramble for money they began walking the very thin line between honest promotion and corruption. In the wave of prosperity and speculation that swept the country in 1835 the directors of the Morris became more interested in their banking privilege than in the canal.

"They embarked on financial ventures which were nothing short of criminal," charges Harlow. "Their bank was a wildcat institution of the worst type. . . . They participated in fake promotion schemes, floated loans on spurious collateral, perpetrated swindles through dummy organizations and divided illegal gains with crooked agents and brokers."

No other historian has gone that far in attacking the Morris Canal and Banking Company, nor can the present writer. Undoubtedly the company engaged in various forms of skullduggery, but the musty court records of Essex County do not reveal that it was ever successfully prosecuted for its alleged offenses.[4]

Though often under attack, somehow the company held together and in 1840 it completed enlargement of the canal, locks and planes, and the extension to Jersey City.

As the boats became larger and the loads they carried heavier, they were cut in two and the ends made watertight so that each half was a boat in itself. Using iron plates and pins they could be quickly coupled together. This device made the strain on the cables and the big revolving drums at Stanhope[5] (fifty-one miles from Newark and the highest point on the canal, 914 feet above the Hudson) much lighter. The trucks (the carriages on which boats were loaded before ascending a plane) were also cut in two. That these innovations were found necessary is better evidence than cold tonnage figures that the Morris Canal was forwarding a tremendous amount of coal, iron and iron ore. But great as it was it could not extricate the company from the web of frenzied finance in which it had become entangled. Finally in 1841 it crashed into bankruptcy. The court-appointed receivers leased the canal to outside parties for three years. They spooned off what cream they could find and wasted no money improving the deteriorating canal.

In 1844 a new Morris Canal and Banking Company was organized with a paid-in capitalization of $1,000,000. The canal bed was examined from end to end and the places where seepage had been occurring for years were lined with clay, as had been necessary on the Delaware and Hudson. To insure a plentiful water supply, two feeders were dug, the one to Lake Hopatcong only half a mile long, the other to Pompton, 3.6 miles. The planes were rebuilt and equipped with new wire cables.

By 1855 the Morris Canal was in better condition than it had ever been. The Morris never had any passenger business, being from first to last strictly a freight carrier. Its tonnage had increased fourfold and totaled upwards of 500,000 tons. By 1860 it amounted to 707,631 tons. These were the most prosperous years the Morris Canal was ever to know as the demand for anthracite was becoming so great that the carriers could not keep up with it. But the Morris was not limited to coal; it forwarded tons of malleable pig iron from the growing number of furnaces in northern New Jersey. In addition new factories were coming into existence, with New York City the natural market for what they produced. This prosperity was regional and should not be construed as marking the beginning of the great industrial complex that metropolitan New Jersey was to become.

By 1857 the railroads were taking dead aim at the Morris Canal. The Morris and Essex Railroad put a branch into Boonton and immediately got a major share of the iron business. Close behind it came the obstreperous Lehigh Valley Railroad, which seemed to enjoy butting heads with any rival that got in its way. By canal, four days were required to move coal from the Delaware River to tidewater at Jersey City; the average time by rail was five hours. No canal could stand up to that sort of competition for long.

Lock 17, Newark, on the Morris Canal. *Courtesy Little Falls, N.J., Public Library.*

Sluice box and waterwheel tower at the top of an inclined plane on the Morris Canal, Montville, New Jersey. *Courtesy Little Falls, N.J., Public Library.*

The dead Morris Canal, October 1924. *Courtesy Little Falls, N.J., Public Library.*

Charcoal burners had been thinning out the once abundant stands of timber at such a rate that some furnaces were forced to close down for lack of charcoal. The invasion of the railroads, bringing anthracite (which made as hot a fire as charcoal) to their door, solved the fuel problem for some foundries. Those that the railroads did not reach had to depend on the canal to keep them going.

The Morris Canal was happy enough to gather these figurative crumbs that fell from the table, but they were not sufficient to keep it prosperous. After 1860 its business steadily declined.

In addition to the Lehigh, the New York and Erie Railroad, the Central Railroad of New Jersey (an amalgamation of smaller roads) and the Delaware and Lackawanna Railroad were busily carving up northern New Jersey to serve their own ends. Although the Lackawanna Railroad had no connection with the Morris Canal until after the dissolution of the latter, it must be mentioned here because it did more to glamorize anthracite than all the other railroads and canals combined. This couldn't have happened without the approval of George Scranton, wealthy iron founder, coal baron and president of the Lackawanna, for whom the city of Scranton was named.

But it must have been with some misgivings that Scranton, who never saw any humor in throwing money away, engaged Earnest Elmo Calkins, the advertising genius of his day, to publicize the virtues of the Lackawanna Railroad and give it "personality" (a Calkins trademark). Presently in the pages of every slick magazine, on streetcar ads and billboards, America was introduced to young, glamorous, beautiful Phoebe Snow, the creation of an artist's fancy. Dressed in white, with her white picture hat, she promptly became the pinup girl of her time. Her picture was always accompanied by a jingle extolling the wisdom of traveling via the Lackawanna. Among children it became a game to see who could quote the most. In the writer's memory, this one was the most popular:

> Says Phoebe Snow
> About to go
> Upon a trip
> To Buffalo:
> "My gown stays white
> From morn till night
> Upon the Road
> Of Anthracite."

As was the case with most of Calkins' promotions, it was successful. Uncounted thousands of tourists and honeymoon couples bound for Buffalo and Niagara Falls made it a point to get there over the rails of the Lackawanna.

In 1871 the Morris Canal, on which $5,100,000 had been expended, was leased to the Lehigh Valley Railroad for 999 years. Under railroad management it continued to deteriorate. In 1903 the Lehigh asked that the state relieve it of the burden of maintaining the Morris. The request was granted and the state took over. Though operating at a loss, New Jersey kept the dying canal functioning twenty-one years. In 1924 the legislature called a halt, decreeing that the old waterway be abandoned. The banks were cut and the water drained out. As if trying to rid itself of a bad memory, the state ordered the destruction of all

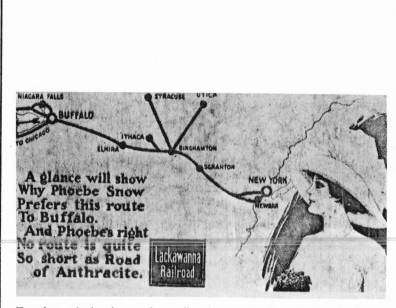

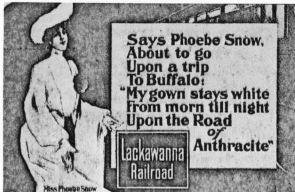

Another Phoebe Snow billboard.

Travelers, whether by canal or railroad, were familiar with Phoebe Snow.

Morris Canal works, even including the senseless dynamiting of the stone viaduct that had carried the canal over the Passaic River at Little Falls.

Today, driving westward on U.S. 202, east of Boonton, you can catch glimpses off to the right of the old canal, down among the weeds. The Morris suffered the final indignity when the city of Newark turned the canal bed into a track for streetcars.

Boats had been leaving the Morris for years before its official demise. Many of them had gone to the Delaware Division of the Pennsylvania state-owned system of artificial waterways. An even greater number had found profitable employment on the combined Delaware and Raritan–Schuylkill Navigation complex, on which, in its most prosperous years, an estimated two thousand boats were competing for business. Such a concentration of canal freighters was never to be seen again on the canals of this or any other country.

Beginning at Easton, the Delaware Division followed the Delaware River southward for sixty miles to tidewater at Bristol, Pennsylvania. If it had been built as an extension of the Lehigh Canal—which many alleged it was—it could not have served the Lehigh better. In 1854, when an outlet lock was built at New Hope on the Pennsylvania side, thirty-five miles below Easton, permitting boats to leave the canal, cross the Delaware River and at Lambertville, on the New Jeresy side, be locked into the twenty-two-mile-long feeder of the Delaware and Raritan mainline, an endless stream of Lehigh anthracite began moving across New Jersey to Raritan Bay and on into New York harbor.

The years that followed were busy ones for the Delaware Division Canal. But the state of Pennsylvania could not make it pay, even though it was doing a tremendous business. Had the Delaware been exclusively a forwarder of coal, which was not the case—it was handling a great amount of timber, lime, charcoal and flour—its failure to show a profit would have been easy to explain. Other divisions of the state's canal system were having the same difficulty. Disgruntled taxpayers charged that the canals were being mismanaged by bungling politicians and demanded that they be turned over to private enterprise.

Added to the growing public discontent was the increasing pressure of the railroads. The result could have been predicted; one division after another of the state canals was sold until only the Delaware Division was left. The Delaware Division was a vital link to the continuing success of the Lehigh Coal and Navigation Company. After being bounced back and forth for a year, the Lehigh got possession in 1866. Half a century later it was still functioning, the last survivor of the network of canals once linked to the Delaware River.

Lock on the Morris and Essex Canal, c. 1900. *Courtesy Edwin P. Alexander Collection.*

11

The Delaware and Raritan

AS EARLY AS 1804 that part of New Jersey over which much of the Revolution had been fought had become the corridor through which thousands of tons of farm produce and manufactured goods were conveyed by wagon from Philadelphia to New York. Except for the tedious and vastly more expensive route by sea, down Delaware Bay and up the open ocean to New York harbor, there was no other way of getting there—no railroads, canals or navigable rivers.

Most of the wagons going back and forth were owned by freighting companies, many of which kept a hundred or more on the road. We have often read about the heavy freighters and bull teams of the pioneer West, but it is doubtful they ever moved the quantity of merchandise that became routine with the wagon companies of Philadelphia, amounting in its most properous years to sixty thousand tons annually. Very few of the wagons employed in this trade were of the traditional Conestoga type; a much lighter vehicle, capable of carrying a five-ton load that could be drawn by a team of sturdy horses, was preferred.

Although the wagon brigades left Philadelphia by different routes, they converged east of Trenton and from there generally followed across the state to New Brunswick what was to become the course of the Delaware and Raritan Canal. With the advent of winter this traffic slackened off or stopped altogether for several months.

Rates for towing on the Delaware and
Raritan Canal, c. 1870. *Courtesy
Edwin P. Alexander Collection.*

Inevitably the extent of the wagon trade convinced a group of farsighted men,
including young Richard Stockton, that a waterway connecting the Delaware with
tidewater in Raritan Bay would capture that business. Calling themselves the New
Jersey Navigation Company, they appealed to the state to grant them a charter
authorizing them to construct a canal "for the purpose of opening communication
between the tidewater of the Delaware and Raritan Rivers," in other words, em-
powering them to construct a canal between the head of the saltwater estuary of
the Raritan at New Brunswick and Bristol, New Jersey, on the Delaware, twenty
miles north of Philadelphia.

Still young and as yet unknown outside his native state, Richard Field Stockton was the scion of a famous and wealthy Princeton family, whose great-grand-
father, as a member of the Continental Congress, had signed the Declaration of
Independence, and whose father was a United States senator. Lustrous as his

lineage was, as Commodore Richard Stockton he was to give it even greater glory when as commander of the U.S. Pacific Fleet, he sailed into San Francisco Bay in 1846 and proclaimed California a territory of the United States.

Just when young Stockton joined the group that had banded together under the name of the New Jersey Navigation Company is uncertain, but for the prestige of his name they made him the head of their organization and he was their spokesman when the state was persuaded to authorize a survey of the proposed canal route in 1816 and again in 1823. But that was as far as the matter got. Sectional differences arose; Pennsylvania owned equal rights with New Jersey in the Delaware River, and when it refused to grant the New Jersey Navigation Company the privilege of diverting the waters of the Delaware for canal purposes, the promotion collapsed, for there was no other sources from which sufficient water could be taken.

But as the years passed and the wagon traffic continued to increase, interest in the canal idea revived. Annually petitions were addressed to the legislature requesting that a commission be appointed to study the feasibility of constructing a canal from the Delaware to New Brunswick. Such a bill was passed in 1829. The commissioners made their report six months later. It was so favorable that negotiations were opened with Pennsylvania to get that state to reverse its position on the use of Delaware River water for canal purposes. This was accomplished, and on February 4, 1830, an act providing for the canal's construction became law. Within hours the Delaware and Raritan Company was chartered. Perhaps by coincidence but very likely by design, the Camden and Amboy Railroad was chartered on the same day.

If there wasn't a secret understanding between the proprietors of the Delaware and Raritan Canal and the owners of the Camden and Amboy Railroad before either had done a lick of work, there soon was one. As projected, they were to pursue a parallel course, the canal running from the saltwater estuary of the Raritan River at New Brunswick to Bordentown on the Delaware; the railroad to lay its tracks from the Amboys (South Amboy and Perth Amboy) on Raritan Bay to Bordentown and down the Delaware to Camden, across the river from Philadelphia, a distance of sixty-one miles as opposed to the forty-four-mile length of the canal.

The canal company immediately sought to charter a railroad of its own from New Brunswick to Bordentown in such close proximity to the canal that a New Brunswick newspaper predicted (facetiously) that the proprietors of the Delaware and Raritan "seemingly intend to lay their rails along the berm bank of the canal. A great saving no doubt if they prevent their trains from toppling into the ditch."

The state could not grant the charter without violating the rights given the Camden and Amboy. A few months later, however, a remarkable consolidation of the supposedly rival companies occurred. Each retained its own identity but pooled its expenditures and revenues.

Construction of the canal began at once, and Canvass White, fresh from his successful completion of the Lehigh, was made chief engineer. With a backlog of $4,000,000 to draw on as needed, the work went forward rapidly on both the main channel and the feeder. With very few locks to be installed, the Delaware and Raritan was a comparatively cheap canal to build. On June 27, 1834, it was opened to traffic, the long feeder which took off from Raven Rock on the Del-

"Horse boat"—one of the last—on the Delaware
and Raritan. *Author's collection.*

aware and joined the canal proper at Trenton having been completed previously.
White built it eighty feet wide, with a depth of eight feet, making it one of the
largest channels ever dug in the United States. But it became so prosperous
that it had to be enlarged some years later.

The Camden and Amboy, which became an important division of the great
Pennsylvania Railroad system in 1870, was as prosperous as the Delaware
and Raritan Canal. Between them they put the once thriving wagon-freighting
trade out of business. Railroad historians never refer to the Camden and Amboy
without recalling that it was Isaac Dripps, the C. & A.'s boss mechanic, who gave
American railroads that distinctive feature, the cow-catcher, and very much as it
is known today. It seems that the road was having no little trouble with derailments caused by livestock getting onto the company's tracks. "Dripps devised a

RATES OF TOLL
ON THE
Delaware and Raritan Canal and Feeder.
1848.

ARTICLES.	WEIGHT.	P. Mile. c. m. r. s	Thro'. cts. r. m. r.	Feeder.	Estimated Weight.	ARTICLES.	WEIGHT.	P. Mile. c. m. r. s	Thro'. cts. r. m. r.	Feeder.	Estimated Weight.
Apples, Peaches, and other Green Fruit,	per bushel,	0 0 4	1½	0 0 4		Marble, Manufactured,	per 1000 lbs.	1 0 0	43	1 0 0	
Bark, if conveyed in Boats,	per cord,	1 5 0	65	1 0 0		Nails,	per ton,	1 0 0	43	0 7 5	
do. do. in Rafts,	do.	3 0 0	1 30	2 0 0		Oysters and other Shell Fish,	do.	0 0 5	2	0 0 5	75 lbs.
Boards, Plank and Scantling, if conveyed						Oyster Shells,	do.	0 0 3	1	0 0 3	40 lbs.
in Boats,	per 1000 feet,	1 0 0	43	0 8 5		Passengers,	each,	2 0 0	86	1 0 0	
do. do. do. in Rafts,	do.	4 0 0	1 75	2 0 0		Posts, Locust or Cedar in Boats,	per 100,	1 2 0	50	0 8 5	?
Brick,	per 1000,	1 2 0	50	1 2 0	4,500 lbs.	do. do. do. in Rafts,	do.	2 5 0	1 00	1 7 0	
Baskets, empty,	do.	1 7 5	75	0 8 5		do. Oak or Chesnut in Boats,	do.	1 2 0	50	0 7 5	
Bran and Shorts,	per bushel,	0 0 3		0 0 3		do. do. do. in Rafts,	do.	2 5 0	1 00	1 5 0	
Burr Blocks,	per 1000 lbs.	0 6 0	25	0 6 0		Paving Stones,	per 1000 lbs.	0 5 0	4	0 2 0	
Cattle,	do.	1 2 0	50	0 8 5		Peas, Green,	per bushel,	0 0 4	1½	0 0 4	
Coal,	per ton,	1 0 0	30	0 5 0		Ploughs,	each,	0 1 7	7	0 1 0	
Corn Meal,	per hhd.,	0 4 7	20	0 4 7	800 lbs.	Plaster of Paris,	per ton,	1 0 0	30	0 5 0	
Charcoal,	per bushel,	0 0 2	1	0 0 2	15 lbs.	Potatoes,	per bushel,	0 0 3	1	0 0 3	
Carriages,	each,	2 5 0		1 7 5							
Carts,	each,	1 2 0	50	0 8 5		Rails, if conveyed in Boats,	per 100,	0 7 0	30	0 5 0	
Clay in Hhds.,	per ton,	0 4 0	16	0 4 0		do. do. in Rafts,	do.	1 4 0	60	1 0 0	
Cider,	per bbl.	0 1 5	6½	0 1 5		Rakes,	per dozen,	0 1 5	6½	0 1 5	
Coaches, Post, and Omnibuses,	each,	3 5 0	1 50	3 5 0	1 ton.	Salt,	do.	0 0 5			56 lbs.
Empty Casks, Hhds.	do.	0 1 5	6½	0 1 5		Shingles, Lathing,	per 1000,	0 5 0	22	0 3 0	
do. Barrels,	do.	0 0 4	1½	0 0 4		Shingles, 18 in. long, in Boats,	do.	0 5 0	22	0 3 0	
Fish, Salted,	per barrel,	0 1 2	5	0 1 2	300 lbs.	do. do. in Rafts,	do.	1 0 0	44	0 6 0	
do. Fresh,	per 1000 lbs.	1 0 0	43	0 8 5		do. 2 feet long, in Boats,	do.	0 6 5	28	0 4 0	
Flax,	per 1000 lbs.	1 2 0	50	0 8 5		do. do. in Rafts,	do.	1 3 0	56	0 8 0	
Flax Seed,	per bushel,	0 0 4	1½	0 0 4	56 lbs.	do. 3 do. in Boats,	do.	1 0 0	43	0 5 0	
Gigs,	each,	1 2 0	50	1 2 0		do. do. in Rafts,	do.	2 0 0	86	1 0 0	
Grass Seed,	per bushel,	0 0 4	1½	0 0 4		Sheep,	each,	0 1 5	6½	0 1 5	
Grind Stones,	per 1000 lbs.	0 5 0	22	0 5 0		Slate, for Roofing,	per 1000 lbs.	0 4 0	16	0 4 0	
Glass,	do.	0 6 0	25	0 6 0		Stone, wrought,	do.	0 7 5	33	0 7 5	3,750 lbs.
Hay or Straw,	per 1000 lbs.	1 7 5	50	0 8 5		Staves and Heading, (bbls.) in Boats,	per 1000,	2 5 0	1 00	1 2 5	
Hogs,	per 1000	3 0 0	1 30	1 5 0	11,200 lbs.	do. do. in Rafts,	do.	5 0 0	2 00	2 5 0	5,600 lbs.
Hoop Poles, Hhd.	do.	1 5 0	65	0 7 5	5,600 lbs.	do. (hhds.) in Boats,	do.	3 2 5	1 40	1 6 0	
do. Bbl.	per barrel,	0 1 5	6½	0 1 5	340 lbs.	do. do. in Rafts,	do.	6 5 0	2 80	3 2 0	
Hydraulic Cement,	per ton,	1 0 0	30	1 0 0		Stone, unwrought,	per perch,	1 0 0	10	1 0 0	2,000 lbs.
Iron, Pig,	do.	1 0 0	30	1 0 0		Timber, in Boats,	per 100 c. feet,	1 5 0	63	1 0 0	
do. Old Scrap,	do.	1 0 0	43	0 7 5		do. in Rafts,	do.	3 0 0	1 00	2 0 0	
do. Cast, and Water Pipes,	do.	1 0 0	43	0 7 5		Vegetables, not enumerated,	per bushel,	0 0 4	1½	0 0 4	
do. Bar,	per bundle,	0 0 3		1 0 0 3		Vinegar,	per barrel,	0 1 5	6½	0 1 5	
Lath, Plastering,	1000 ft. r. m	0 5 0	22	0 5 0	100 pieces.	Wagons, 2 horse,	each,	1 7 5	75	0 7 5	
do. Shingling,	per 1000 lbs.	0 2 0	8	0 2 0		do. 1 horse,	do.	0 8 6	37	0 8 6	
Lime, Sand, Manure and Iron Ore,	per 1000 lbs.	0 2 0	8	0 2 0		Wood, Hickory or Oak,	per cord,	1 0 0	43	1 0 0	
Marble, Unwrought,	do.	0 5 0	20	0 5 0		Wood, Pine,	do.	0 7 5	33	0 7 5	
Marble, Sawed,	do.	0 7 5	33	0 7 5							

All other articles, not enumerated, at four cents per ton per mile, if carried through; and one cent per ton per mile, from or to Easton, or if landed or taken from any point on the line of the canal.

Coal passing through canal by way of feeder. From the head of the feeder including the lock on the Jersey side to New Brunswick, 35 cts. per ton. For way tolls ½ cent on feeder, 1 cent on main canal per ton per mile, and one dollar a boat for passing lock on Jersey side.

A drawback of 10 cents, allowed, on coal going north of Manhattan Island, and east of Hurl Gate.

On all vessels or boats regularly employed in the transportation of freight through the canal, four cents per mile, exclusive of cargo, and four cents for passing each lock.

On all transient vessels or boats which may pass through the canal with over 30 tons of cargo, four cents per mile, exclusive of cargo, and four cents for passing each lock; with less than 30 tons of cargo, twelve and a half cents per mile, exclusive of cargo.

On all vessels of less tonnage than 30 tons, which may have gone outside, and returning through the canal, twelve and a half cents per mile, exclusive of cargo, and for each additional ton or tonnage over 30, half a cent per mile, exclusive of cargo.

All coal vessels or boats having full freight, will not be required to pay mileage on the vessel or boat, such vessel or boat having paid toll for full cargo when passing the canal, will, on returning empty, be permitted to pass toll free.

All vessels or boats passing the tide lock at New Brunswick shall pay toll on the vessel and cargo, equal to one mile, except three-fourths of an hour before and after high tide, at which time they may pass free.

In either of the following cases, the company reserve the right to charge the highest tolls authorized by their act of incorporation, to wit: four cents per ton per mile for every species of property carried on the canal; and two cents per ton per mile on the feeder; when the rules and regulations published by the company shall not be complied with, when boats, or vessels, shall sink or get fast in such positions as to obstruct the navigation, or when the vessel, or boat, shall be towed by teams not belonging to the owner of such boat, or vessel, not regularly employed in towing under some special arrangement with the company.

Toll rates on the Delaware and Raritan Canal, c. 1848. *Courtesy Edwin P. Alexander Collection.*

low truck and attached it to the front end of the engine. Sticking out ahead of the two truck wheels were two long and pointed bars of wrought iron. 'That rig,' Isaac Dripps declared of the formidable weapon 'ought to impale any animal that may be struck and prevent it from falling under the engine wheels.' " [1]

It worked only too well. A wandering bull got in the path of the locomotive and was speared so thoroughly that block and tackle were required to free it. Dripps modified his invention, replacing the prongs with a heavy iron bar placed at right angles to the track. Strangely, the pilot (as railroaders refer to it) which, in addition to preventing many derailments has saved many human lives, has come into use on very few foreign railroads.

The Delaware and Raritan could not have come upon the scene at a more fortuitous time. The great acceptance of anthracite as a preferred fuel was in full

State Street lock on the Delaware and Raritan Canal at Trenton. *Courtesy Newark Public Library Collection.*

swing. The problem was to find boats in which to move it to market. They were being built as quickly as they could be knocked together. But there were never enough. As early as 1841, 119,972 tons of anthracite reached Raritan Bay via the canal, in addition to which it forwarded up to 30,000 tons of burnt lime and other products. Having no passenger business (which would have been a nuisance) enabled the canal owners to concentrate their attention on increasing the flow of anthracite from the Pottsville region on the Schuylkill and subsidizing boat builders, offering prizes for greater speed in turning out craft. They demanded bigger boats of not less than eighty tons burden, the limit that could be handled on the Schuylkill Navigation. Matters were expedited when loaded boats coming off the Schuylkill were herded together on reaching Philadelphia and towed up the Delaware by steam in flotillas of fifteen or more to Bordentown, where they entered the Delaware and Raritan Canal.

By 1844 it was obvious that if the canal was to continue to increase its business it would have to be enlarged again. Not only would that be costly, but construction would seriously curtail traffic until it was completed. However, it was decided to make the improvement. Work got under way the following year and was completed twelve months later, no small feat. What it cost was not disclosed, which was in line with company policy never to take the public into its confidencce. But it was widely rumored that the D. & R. was pinched for money. In view of the lush years it had enjoyed, that seemed hardly possible.

Between them, the Delaware and Raritan Canal and the Camden and Amboy Railroad had, by the terms of their charters, a guaranteed monopoly of the right to transport freight and passengers across the state of New Jersey from Philadelphia to New York. It had been charged for years that the two companies, in essence really one, had paid dearly for the privileges they exercised. As drawn, the canal charter guaranteed that no competing canal could be built within five miles of its line. The C. & A. charter went even further, stating that no new railroad could be built across New Jersey without the consent of the combined companies.

As these giveaways became known it could not be doubted that they had been pushed through the legislature by corrupt politicians. The scandal became ugly and it was alleged that the "Trust" (the twin transportation companies) had paid out in graft as much as $2,000,000 over the years.[2]

The so-called Trust ignored the outcry leveled against it and went about its business. It paid its taxes, made sketchy reports to the state, as required by law, declared modest dividends to its stockholders, but disclosed nothing about the disposal of its net income.

In 1854 its long feeder was connected with the Delaware Division Canal at Lambertville, as previously stated, which gave it practically all of the business of the Lehigh Canal. The aggressive Philadelphia and Reading Railway had laid down the tracks for the Schuylkill to Pottsville extension. The Schuylkill Navigation felt that competition at once, but it wisely adapted itself to the circumstances. It owned several thousand dump cars and it entered into an arrangement with the Reading to deliver coal from the mines to the water's edge, over its own narrow-gauge tracks. This was no trifling business for, by one means or another, the annual gross tonnage handled by the Navigation between 1855 and 1867 fell below 1,000,000 tons in only two years. Most of it reached New York via the Delaware and Raritan. In 1860, out of more than 1,200,000 tons of anthracite forwarded by the D. & R., almost 900,000 tons came from the Schuylkill Navigation and only 300,000 from the Lehigh.

To stay with the Navigation a minute longer: It had led a charmed life. It had been enlarged twice, new and higher dams had been built and a total three miles of its channel was riprapped (paved) with stone to lessen seepage. When its troubles came, they came in a cluster. They began with a miners' strike that lasted for several months in 1869. The strike was followed by a prolonged drouth. Philadelphia, threatened with a water famine, was compelled to lower the water level above its Fairmount Dam so drastically that boats grounded and could not be moved. Hard on the heels of the drouth came the worst floods that ever ravaged the Schuylkill Valley, destroying the canal channel and wrecking the dams.

These events dealt the Schuylkill Navigation a fatal blow. Unable to continue, it turned over to the Reading Railway, on the usual 999-year lease, its rights, title and property. The Reading spent a vast amount of money repairing the canal, but it was barely back in commission when the Pennsylvania Railroad Company, which was in the process of gobbling up small roads and welding them together into the great Pennsylvania System it was to become, gained control of the Camden and Amboy in 1871, and along with it the D. & R. Canal. In an exhibition of business savagery not yet common in the United States, it decreed that coal from the Schuylkill mines would not be permitted to pass through the Delaware and Raritan. Needless to say, the purpose of this move was to cripple the Reading, which controlled the Schuylkill mines.

It was typical of what the opposition came to expect from the Pennsylvania, which spawned a line of rugged individualists to whom its advancement became a

Elysian Fields, Hoboken,
graveyard of canal boats.

religion. Men like John Edgar Thomson, for instance, one of the greatest all-around railroad men America ever produced. He came to the Pennsy, a lowly assistant civil engineer, via the Camden and Amboy, and began his rise to the top. "He was at times a difficult man, taciturn, abrupt to rudeness, and given to doing startling things without consulting his directors." [3] But when he died in 1874, he left behind him a railroad property that was then, and still is, the foremost carrier in this country. There were others in the Thomson tradition—Tom Scott, George Roberts, Alexander Johnston Cassatt, Samuel Rea. They fought with tooth and claw, suborned judges, bought politicians and agreed with William H. Vanderbilt's ukase that "the public be damned." [4]

Being ruled off the Delaware and Raritan hurt the Reading, but it was too strong to be put out of business in that fashion. Extending its trackage from Trenton to Bound Brook, it made connection with the Central of New Jersey and began shipping coal to the northern and eastern markets exclusively by rail. As a consequence, the deteriorating Schuylkill Navigation, long so prosperous, slowly expired.

Striking back, the Reading cut its freight rates, forcing the Pennsylvania to do likewise. The latter then reduced canal tolls, although the exclusion of Schuylkill coal was costing the D. & R. a million tons a year. If the lessees saw any future for the canal and expected it to operate at a profit, they went about it in ways that pass understanding. At great expense they installed a number of swing bridges, steam-operated. On the other hand, some years later, they permitted the building of low bridges across the once vital feeder, closing it to the passage of canal boats. Of course, by then the Lehigh Valley Railroad had garnered most of the coal being mined in the Lehigh Valley, leaving little to be sent down the Delaware Division to the Lambertville feeder. But that does not explain the seeming intransigence of the Pennsylvania.

The fate that had pushed other canals into oblivion surely awaited the Delaware and Raritan. But it was to be a long time coming. By the turn of the century, the horse-drawn boats had all but disappeared from the D. & R. Steam tugs and gas boats now towed the barges across New Jersey. In the early 1930s it was still a common sight to see them being herded up the Hudson to the coal docks on the Jersey side of the river.

This was before the river became polluted and the harbor was crowded with small craft and steamers to Coney Island, the Jersey shore and Long Island points. Other New Yorkers, needing a bit of "country" but with less money to spend, compromised by crowding aboard the Hoboken ferry on holidays and weekends and crossing the river to Hoboken's Elysian Fields, where children could romp and play while their elders lounged under the trees until it was time to open the lunch baskets.

Along with the canal those bucolic pleasures were passing. Almost before New York noticed, they were gone. The once majestic Hudson was becoming befouled. So was Elysian Fields. As *Harper's Weekly* bothered to notice:

"Hoboken's Elysian Fields, once a happy picnic ground, is becoming a grave-yard for decaying canal boats, with squatters finding shelters in the decaying hulks until wreckers come to break up the soggy timbers." [5]

But the economy was strong and the country was taking gigantic steps forward. For instance, in Jersey City no fewer than a dozen competing railroads were waiting to take a traveler almost anywhere he cared to go. It was said that the Pennsylvania was even thinking of digging a tunnel under the Hudson so that its trains, operating by electricity, could arrive and depart from the heart of New York City.

The New York Times predicted that the proposed undertaking "will be one of the great engineering feats of the century." To the few remaining horse-boat captains on the Delaware and Raritan, it meant that the inevitable end of their occupation and way of life was edging a little nearer.

12

The State-Owned Canals
of Pennsylvania

ALTHOUGH WE MAY KNOW BETTER, we are inclined to think of the Susquehanna as strictly a Pennsylvania river, ignoring the fact that when the north branch meets the west fork at Northumberland it has pursued a curving, twisting southerly course for more than 260 miles from Cooperstown, New York, and Otsego Lake, its source. Strong with the combined flow of the west branch and the north fork, it continues south-southeast from Northumberland, gaining additional strength from the Juniata. Sweeping by Harrisburg, the great river hurries on past Peach Bottom, the last of the Pennsylvania towns, which is remembered for its vanished glory, and in Conowingo Lake crosses the invisible Mason-Dixon line into Maryland waters and Chesapeake Bay, a total journey of nearly five hundred miles.

Until the arrival of Charles Williamson, one of America's early super-salesmen, in the Genesee Valley in the early 1790s, the upper Susquehanna was regarded as navigable only by canoe and keelboat. A dashing figure and former captain of British troops, Williamson had been captured on the high seas during the Revolution and quartered for the duration in a Roxbury, Massachusetts, home, where he had become so Americanized that he married the daughter of the household. He was in the Genesee Valley as the agent of Robert Morris, the land speculator, to promote the sale and development of thousands of acres of rich

Double coal boats on the Pennsylvania canals along the Susquehanna River.

The two great branches of the Susquehanna River at Northumberland. *Courtesy National Archives.*

farmlands. Folklorists describe him as a tall, handsome man, regarded with awe by country bumpkins and backwoodsmen as he raced about on his blooded horse, his long cape floating out behind him.

He soon discarded his original scheme of making the Genesee Valley into a region of landed estates for the very wealthy, such as existed in England. To attract humbler men, he staged fairs, racing meets, athletic contests and the like. He touched solid ground when he centered his attention on the Susquehanna and declared that with minor improvements it could be made navigable all the way to Chesapeake Bay, thus opening a great market for Genesee Valley produce. To prove it he designed that ugly but serviceable flat-bottomed creation, the ark.[1]

Other than the canal around Conewago Falls little had been done to improve the navigation of the Susquehanna. Hundreds of arks—scows seventy-five feet long and sixteen feet wide, pointed at both ends to make them less difficult to handle, and controlled (hopefully) by long pine sweeps mounted on sternposts several feet high—ran the river in the freshets. Many of them crashed on rocks, had their bottoms ripped out, and swamped, their cargoes lost. As on the Lehigh, it was impossible to work them back upstream. Cutting them up into lumber became a steady business at Port Deposit and Havre de Grace.

The Susquehanna formed a natural dividing line between the eastern third of the state and the country to the west. The prime interests of the two sections were widely divergent. Philadelphia controlled the business of the eastern division and was intent on capturing the trade of the west, centering on Pittsburgh, which was booming with the heavy influx of settlers taking up land in the new states of Ohio, Michigan, Indiana and Illinois. The Susquehanna Valley, on the other hand, was interested only in winning a larger share of the Philadelphia market. For years these opposing interests made it impossible to get any program of internal improvements through the legislature, and the impasse was not broken until the state bailed out the bankrupt Union Canal Company.

As previously noted, the Union Canal was deluged with business almost from the day it went into operation. But the blunders made in constructing it—small locks, narrow channel and failure to provide an adequate water supply—ate up the profits faster than they could be earned. Although an additional $4,000,000 was spent on the Union, it had to be written off as a failure.

Canal enthusiasts suffered a setback in 1820 when the legislature authorized construction of a state-owned and operated turnpike extending westward across Pennsylvania from Wrightsville on the Susquehanna to "the conjunction of the Monongahela and Allegheny rivers"—in other words, Pittsburgh.

This was the original Pennsylvania Turnpike, which reached the Susquehanna largely by way of the Three Mountain Trail, sections of which mark the route of today's Pennsylvania Turnpike. In the Enabling Act no provision was made for building or improving the approximately eighty-two miles of road between Philadelphia and Columbia (across the river from Wrightsville), since the excellent Lancaster Road, the first surfaced highway in the United States, could be used as far as Lancaster, where it connected with the toll road of the Lancaster Turnpike Company which covered the remaining miles to the Susquehanna and the established Wrightsville Ferry.

The river is very wide at that point; the combination wooden bridge and aqueduct that was built there in the 1820s was a mile and quarter long. John Wright was not the first man to establish ferry service at that crossing, but he was the first to be granted a royal patent. At his death in 1766 he had been conveying

A time schedule for towing packet boats on the Pennsylvania Canal, 1840. *Courtesy Edwin P. Alexander Collection.*

travelers back and forth across the Susquehanna for more than half a century. He made his headquarters on the east bank, where in time he had four or five neighbors. The settlement was too small to have a name. Across the river, where Wrightsville now stands, there were no permanent inhabitants.

After his death his nephew, Samuel Wright, carried on the ferry business. As the settlement grew, it became known as Wright's Ferry and remained so until the spring of 1789 when it appeared that the Congress might be persuaded to establish the national capital in Pennsylvania. Wright made a gift of ten acres of land to the government and changed the name of Wright's Ferry to the more patriotic one of Columbia. Although the advocates of Pennsylvania as the national capital failed to capture the prize, Columbia experienced a boom. Enterprising Samuel Wright plotted a townsite and sold his lots by lottery at fifteen shillings a chance, the winners receiving a lot for nothing. This novel exploitation attracted home-seekers as well as speculators. Charmed by its natural advantages and the wide vistas of the majestic river, hundreds of newcomers and the wide vistas of the majestic river, hundreds of newcomers settled in Columbia and prepared it for the role it was to play in the unfolding history of Pennsylvania. Across the Susquehanna the town of Wrightsville became established. Since the Susquehanna would have to be bridged to accommodate the traffic the turnpike would generate, the Columbia-Wrightsville crossing appeared to be the most advantageous place. So it was not without reason that the surveyors began driving their stakes westward from there two years later.

Figures regarding the tonnage carried by the various canals are readily available. But we know little or nothing about the amount of business done by the early American turnpikes, of which the Pennsylvania Turnpike was the most important. Some years ago, when relating the story of the Three Mountain Trail, the author remembers the difficulty he had in ascertaining reasonably accurate figures covering the number of cattle, horses, mules, hogs and even turkeys that were "walked" across the Pennsylvania mountains to the great market at Harrisburg. It is therefore a pleasure to be able to present the amount of business done by the Pennsylvania Turnpike in 1832, three years after its completion:

Broad wheel wagons	6,359
Light vehicles	3,110
Cattle	6,457
Sheep	2,853
Hogs	40
Draft and riding horses	23,381
Total number of tolls	42,200

The amount of toll collected depended on how many stations were passed. Tolls were high—excessively high was the universal complaint. Of course, the Pennsylvania Turnpike suffered from the same ills that beset the National Road down in Maryland—heaving frosts, floods and washouts. The mountain division could not be kept open in winter nor could the rising annual deficit be curbed. From the beginning when legislation for building the turnpike was under consideration, there was a great body of opinion that claimed the turnpike would not solve the state's transportation problem. They had not been able to prevent the passage of the legislation but they had not been silenced. In December, 1823, the legislature appointed a committee to "study the matter of building a canal from the Susquehanna to the course of the Ohio River" (again, Pittsburgh). The findings were so favorable that a board of commissioners was named on April 1, 1824, to "explore the question further, hold hearings and report its findings to the spring 1825 session of this body."

In January, 1825, there occurred in Philadelphia a mass meeting of several hundred delegates, legislators and prominent citizens from various parts of the state, including Pittsburgh. They were there to voice their unqualified support of legislation that would put Pennsylvania squarely in the canal business as builder, owner and operator. Nicholas Biddle, the Philadelphia financier and president of the Bank of the United States, was one of the speakers. Pennsylvania, said he, could not hope to retain its high commercial rank if it continued to lag behind New York State and Maryland in opening communication by water with the West.

That was what the assemblage wanted to hear. Other speakers sounded an even more doleful note. The most extravagant demanded that the state build a grand canal connecting Philadelphia with the Ohio and Lake Erie. To many that seemed to be asking for too much, but in the end such a program did not fall far short of being achieved.

In February, 1826, a so-called Canal Convention gathered at Harrisburg and the intimidated legislature passed a bill calling for the construction of a canal "connecting the Susquehanna with the Allegheny River." A few months later, with appropriate ceremonies, ground was broken at Harrisburg on July 4.

There being nothing in the existing legislation about connecting the Susquehanna with Philadelphia, Columbia became the recognized eastern terminus of

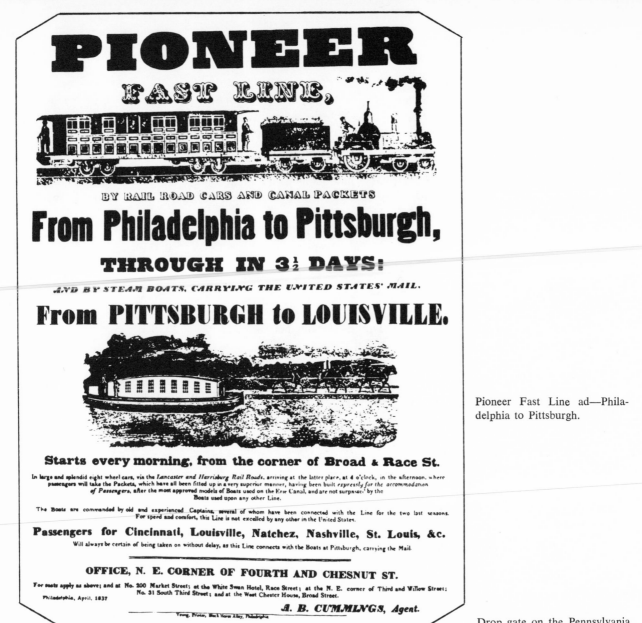

Pioneer Fast Line ad—Philadelphia to Pittsburgh.

Drop gate on the Pennsylvania Main Line Canal. A big improvement on the old swing gate.

Sectional canal boat leaving Broad Street, Philadelphia, for its long journey to Pittsburgh.
Courtesy Library of Congress.

the Pennsylvania Main Line Canal. James Geddes and Nathan Roberts were brought in from the Erie to direct construction. Later, Canvass White joined them. Work began on both ends of the Main Line and proceeded without serious mishap.

Long study had convinced Geddes and Roberts that an all-water route was impossible, or at least too costly to be contemplated. The shortest and most feasible course they could devise was: from the east, to go up the Juniata River from the Susquehanna to Hollidaysburg at the foot of the Alleghenies; and in the west, to follow the Allegheny River as far as the Conemaugh, and turn up that mountain stream to the base of the mountain barrier, which reached an elevation of 2,291 feet. This left a gap of 26½ miles between the Pittsburgh and Juniata divisions. Geddes and Roberts proposed to close it by means of a railroad—really a series of inclined planes.

Work on the western or Pittsburgh Division advanced so rapidly that it was opened as far as Johnstown in 1829. A few months later part of the Juniata Division welcomed its first boats. That same year the turnpike was completed. Although its course was far to the south of the Main Line, it facilitated the bringing in of heavy material for the canal builders.

As might have been expected, taxpayers in the tier of counties bordering New York State, especially in the northeastern section, objected violently to having to pay for building a canal that would not benefit them. The only way they could be placated was for the state to construct a system of branch canals. Over the objections of Philadelphia interests, which protested that the Main Line Canal should be finished before any new work was undertaken, four lateral canals were authorized. They were surveyed in 1827 and excavation began the following year. Three of the four were important. The Susquehanna Division left the Main Line at the junction of the Juniata and the Susquehanna and continued up

Sectional canal boat being drawn up the inclined plane of the Portage Railroad at Hollidays-burg. *Courtesy National Archives.*

the latter to Northumberland, a matter of forty miles, where it forked: one branch continued up the Susquehanna to Wilkes-Barre and beyond, heading for the New York State line and a connection with the Chemung and Chenango canals of that state; the other fork followed the west branch of the Susquehanna to Williams-port. The third so-called branch canal, which was independent of the others, was the Delaware Division. It began at Easton and followed the Delaware down to tidewater at Bristol. A fourth canal, the twelve-mile Wiconisco, was also author-ized.

Philadelphia, not to be outdone, pushed through legislation for construction of a railroad connecting that city with Columbia on the Susquehanna and the Grand Canal. This primitive railroad, operating on strap-iron tracks, its tiny cars moved by horses, was completed in 1834. Four years later its first locomotive chugged into Columbia.

In 1832 the Susquehanna Division was completed to Hollidaysburg at the foot of the Alleghenies, 172 miles from Columbia. Although in that distance a boat had to pass through 108 locks, which exceeded the number on the entire length of the Erie, travelers appear to have made good time by using the railroad between Philadelphia and Columbia and transferring at that point to one of Leech and Company's fast packets, or vice versa if journeying in the opposite direction. If they were westbound, they spent the first night out at Lancaster, reached Columbia late the next afternoon and arrived at Hollidaysburg on the evening of the third day. Leech and Company made a practice of keeping their horses on the trot, which undoubtedly was hard on the animals, but the time saved was appreciated by passengers.

Although canal engineers accomplished far greater feats than the double chain of locks at Lockport on the Erie and the Portage Railroad across the Alleghenies, these are the most often mentioned. You will recall that the inclined plane in connection with a canal was first used at South Hadley Falls in Massachu-setts. In that instance the power by which the boats were raised was waterpower. Benjamin Wright used the same principle when he built the planes for the Dela-ware and Hudson Canal Company. When James Renwick installed a series of inclined planes on the Morris Canal he demonstrated that canal boats could be lifted to the desired elevation by a series of connected planes, and with far less strain on the engines that supplied the motive power than by endeavoring to make the ascent in one bite. All of this was known to Geddes and Roberts when they set about building the Portage Railroad. Certainly Canvass White was aware of what Dr. Wright and Dr. Renwick had accomplished. When he was brought in as consulting engineer there can be little doubt that all three were in agreement on how to get the Main Line over the Alleghenies, but instead of building a series of connected planes, as Renwick had done, they decided on an innovation: they would build separate planes, and by means of horse-drawn cars running on rails, move freight and passengers from the head of one plane to the foot of the next.

It was a practical idea, and it worked. On the western slope they found it necessary to bore a nine-hundred-foot tunnel at Staple Bend on the Conemaugh, four miles east of Johnstown. The rest was comparatively easy. At Pittsburgh they created a slack-water pool on the Allegheny and by means of a viaduct crossed the town to the Monongahela basin, about where today's Pennsylvania Railroad station stands.[2]

The total trackage of the Portage Railroad added up to slightly more than 37 miles: 10.1 miles on the eastern slope and 26.5 miles on the western. Until it was double-tracked, so that east and westbound cars could pass one another freely, traffic jams were unavoidable.[3]

In 1834 the Main Line Canal was officially opened from end to end—606 miles from Columbia to Pittsburgh, the longest and most costly undertaking of its kind in America. It had cost $10,338,133. The important branch canals had already come into operation, and with the exception of the Delaware Division, were prosperous. Work on the extension connecting Pittsburgh with Lake Erie, which was never to be completed, was under way. Pennsylvania could be rightfully proud of its system of canals and the turnpike. It had gone heavily into debt to

Boats about to ascend the lower plane of the Portage Railroad.

The inn at Hollidaysburg on the Portage Railroad.

Sectional boat ascending the Portage Railroad.

►

Old stone sleepers of the Portage Railroad.

build these improvements, but it was believed that they would pay for themselves as well as be a source of revenue for the state for many years.

While the Main Line Canal took no great amount of business away from the turnpike, it had a dynamic effect on the commodity markets of the Ohio River towns and throughout the southern half of Ohio. Flour that had been selling for $4.00 a barrel in Cincinnati in 1826 brought double that figure in 1835; wheat rose from $.25 a bushel to $1.00; whisky climbed from $.18 a gallon to $1.50. Every steamboat that chugged up the river to Pittsburgh bore farm commodities of one kind or another. Meanwhile more and more boats appeared on the canal, burdened with machinery and merchandise from populous eastern Pennsylvania.

Fast packets, carrying only first-class passengers, strove to maintain a six-day schedule between Philadelphia and Pittsburgh and usually managed to keep it. This equaled the time consumed in going from Albany to Buffalo on the Erie. Emigrants bound west to take up land in Indiana and Illinois were transported on slower boats at a lower fare.

A number of celebrated English visitors to this country had occasion to travel by canal boat. Without exception they wrote caustic accounts of the inconveniences and discomforts they were forced to undergo. None more so than Charles Dickens, the famous author. Later on it will be a pleasure to acquaint the reader with some of his more critical observations. It is timely, however, to inject here his opinion of Pittsburgh. "Pittsburgh is like Birmingham—at least its townspeople think so. It is distinguished by the great cloud of smoke hanging over it. The similarity ends there." [4]

Coming on the heels of two wild years of prosperity and reckless speculation, the panic of 1837 rocked the nation on its heels. Several states were unable to pay the interest due on their bonds; Indiana defaulted and Illinois repudiated its entire debt. In the East, Pennsylvania found itself in dire straits. Her canal system was bringing in $700,000 yearly in tolls, but in building it she had amassed a debt of $24,000,000, on which the tolls were not sufficient to pay the interest. Lotteries had recently been outlawed, so there was no help from that direction, but by juggling funds that had been earmarked for education, Pennsylvania weathered the storm temporarily. All construction work was halted. But despite such economies the state's debt continued to rise. The panic of '37 was followed by the severe depression of 1839. Pennsylvania was hard hit again. While Governor Porter viewed the future with confidence and insisted that the Portage Railroad be replaced by a tunnel through the mountains, the state for the first time in its history could not meet its obligations when they fell due in February, 1840.

The disintegration of the state-owned system of canals was at hand. Rising sentiment in favor of selling them to private interests developed. When the question was put to the voters it was decided in the affirmative by a substantial majority. No buyers appeared until 1844, when the Pennsylvania Canal and Railroad Company (unrelated to the Pennsylvania Railroad Company) offered $20,000,000 for the property. It was accepted, but the company could not dispose of its stock and the sale was not consummated.

The branch canals had responded to the reviving economy of the nation and were doing a vast amount of business, but keeping them in condition was costing the state more than they were bringing in. One after another they were sold off, the state retaining the Main Line. But warning of its demise became apparent when

A Leech fast packet for Pittsburgh—horses on the gallop.

Train on the old Columbia Railroad passing through Lancaster at street level. *Courtesy National Archives.*

the Pennsylvania Railroad was chartered in 1846 to build a railroad from Philadelphia to Pittsburgh.

The railroad company had the audacity to parallel the Main Line by building up the Juniata and down the Conemaugh. It put its first train into Pittsburgh on December 10, 1852—fifteen hours from Philadelphia. It was incredible! No canal could contend with competition of that sort.

The Pennsy delivered the final blow when in 1857 it bought the Main Line at its own price and relegated it to the role of a leisurely freight carrier.

13

The Susquehanna and Tidewater

IN CANAL CONSTRUCTION it was the invariable rule that unforeseen difficulties would have to be faced, resulting in skyrocketing costs. They could be anticipated but they could not be measured until encountered. The wonder is that the handful of men who came to be recognized as the foremost engineers of their time succeeded as well as they did. They brought to the job no previous experience and knew little or nothing about hydraulics. Prior to working on the Erie Canal, Dr. Benjamin Wright's only engineering credentials were that he had been a part-time surveyor. Jervis, Roberts, Geddes and several others including young Canvass White were no better qualified. White, the genius of the lot, took a year's leave of absence and went to Europe in 1817 to study the canals of England and the Low Countries, tramping a thousand miles of towpaths in his quest for knowledge.

It is not surprising that the estimated cost of constructing this or that canal invariably proved to be ridiculously low, usually due to ignorance. These were honest errors, quite apart from those instances previously mentioned when unscrupulous promoters published misleading statements of estimated costs to stimulate the sale of stock.

If the public had known what certain canals were to cost before being completed, it is doubtful if they would have been constructed. The losers in such instances were the investors. But any region a canal reached was the better because of it. It opened communications between hitherto remote districts, stimulated the growth of towns and provided the farmer with a market. Out in thinly populated Ohio, where the rich soil produced bountiful quantities of corn and hard grains, the farmer was literally living in poverty in the midst of plenty; on the farmstead the price of wheat and shelled corn, when they could be sold at all, was twenty-five cents a bushel. Shortly after the completion of the Ohio and Erie Canal the price at canal side rose to a dollar. It became common practice for farmers to load their wagons and drive fifty miles or more to the canal.

Being widely separated geographically, canals seldom came into competition with one another. Though much has been said about their contribution to the public good and the welfare of the communities they reached, they were financed and constructed for the basic purpose of encouraging a flow of trade to their respective terminuses. Few major canals crossed state lines. When one did, the legislature of the state about to be invaded jealously guarded the rights of its constituency, even, as in the case of the Susquehanna and Tidewater Canal, refusing to grant a charter lest it should drain business away from Pennsylvania to Maryland. The extent of that business was measured by the widely esteemed *Niles' Weekly Register*, which reported that at the end of the 1829 boating season "a count made at Harrisburg shows that over a thousand rafts, 236 arks and several hundred keel boats have passed down the Susquehanna this year."

Financed by Baltimore money, the proposed canal, forty-three miles in length, was to extend from Havre de Grace northward across the Maryland-Pennsylvania line, up the Susquehanna to Wrightsville and a connection with the Main Line Canal. Maryland had given a charter to the Tidewater Canal Company, but for years the Pennsylvania legislature had refused to grant one. Leading the opposition was the bloc of legislators from Delaware and Philadelphia counties who were controlled by the merchants, bankers and manufacturers of the Quaker City. On the other side of the controversy were the southern counties along the Pennsylvania-Maryland border who were not only weary of being dominated by the Philadelphians but who also saw some advantage to themselves in a canal that would put them in touch with the extensive Baltimore market.

After three years of bickering the matter was finally settled by the granting of a charter to the Susquehanna and Tidewater Canal Company. This came about largely because it had occurred to some of Philadelphia's business leaders, belatedly to be sure, that the trade brought down the proposed canal need not necessarily go to Baltimore; on reaching salt water, tugs could tow flotillas of canal boats across the bay to the Chesapeake and Delaware Canal and up the Delaware River to Philadelphia wharves.

Time proved this to be a sound conclusion. Fully one-third of the boats plying the Susquehanna and Tidewater found their way through the Chesapeake and Delaware Canal to Philadelphia, where they discharged their freight and took on cargo for the return voyage. Although the boats employed in this trade had an average carrying capacity of seventy-five tons, they were open boats and not seaworthy. Consequently whenever Chesapeake Bay was on a rampage, traffic was interrupted, for there was twelve miles of open water between Havre de Grace and the entrance to the Chesapeake and Delaware Canal.

At Wrightsville on July 4, 1836, with Governor Veazey of Maryland and Joseph Ritner, the anti-Masonic Governor of Pennsylvania, on hand with their staffs to give the occasion an aura of importance, construction of the Susquehanna and Tidewater Canal got under way. Its charter authorized the company to build a canal from Wrightsville southward along the west bank of the Susquehanna to the state line. It was also empowered to erect a dam across the Susquehanna from Wrightsville to Columbia, on the east bank "to create a pool across which boats could freely move back and forth."

You will recall that Columbia was the terminus of the Philadelphia and Columbia Railroad. The importance to the canal of making connection with the railroad is understandable—it made only one transshipment of freight necessary.

The managers of the canal company were so confident of success that they spent money with a lavish hand, excavating a channel with a surface width of fifty feet and depth of six feet. The locks they installed were 170 feet long and 17 feet wide—large enough to accommodate boats of 150 tons capacity. While not a long canal, its builders meant it to be one of the finest ever constructed in the United States. Certainly it proved to be one of the most expensive ever dug, the net cost being $80,000 per mile.

Construction proceeded rapidly, and in October, 1839, water was turned into the channel. A great celebration greeted the first boat to reach Havre de Grace. A week later what was regarded at first as the usual autumnal rains began to lash the Susquehanna Valley. But the deluge continued day after day. The canal began to suffer as breaks occurred in the not yet fully settled berm and towpath. Below Conowingo Lake, the dam and costly viaduct were partly destroyed. The canal was so severely damaged that it was out of commission for a year, the repairs costing almost a million dollars. But it slowly recovered and two years later was handling about as much business as could be accommodated, the bulk coming from the Pennsylvania Main Line. Most of the tonnage was coal.[1] But a large part was dressed lumber, farm products and the output of small-town factories.

During the late 1840s upper Chesapeake Bay must have presented a busy scene, with tugs towing fleets of squat canal boats down the bay from Havre de Grace to Baltimore, and other tugs pushing or pulling similar freight carriers from Turkey Point and the Chesapeake and Delaware Canal to the Susquehanna and Tidewater. The amount of business the latter was doing was deceiving, for the S. & T. was only a short-haul canal, really just a forwarder of freight that originated or was destined for points long distances from its terminus at Wrightsville, beyond which the canal did not share in the tolls. Necessarily the company had to operate on a thin margin of profit, but the volume of traffic handled was great enough to keep it prosperous for twenty years.

The old Union Canal, always in difficulties, had received a mortal blow when the state of Pennsylvania granted the Susquehanna and Tidewater a charter.

Completion of the Lebanon Valley Railroad from Reading to Harrisburg in 1857 added to its woes, and the great flood of 1862, which destroyed the Swatara dam and led to the abandonment of the Pine Grove branch, put the Union in bankruptcy. The Philadelphia & Reading acquired the property at a sheriff's sale.

The demise of the Union Canal had no great effect on the Susquehanna and Tidewater, but the building of the Baltimore and Susquehanna Railroad, which paralleled it from Havre de Grace to Wrightsville, did. The competition cut deeply into the canal's Baltimore business.[2] A few months later the first shot was fired

at Sumter and the long, bitter Civil War began. Soon, both railroad and canal were confronted with regiments of Pennsylvania volunteers hurrying to the front and mountains of freight that had to be moved at once.

After the first few weeks, however, battle lines became stabilized and the war appeared to be far removed from the Susquehanna. Then, early in 1863, General Lee and the Army of Northern Virginia surged into Pennsylvania. On June 28, two days before the crucial Battle of Gettysburg, General Jubal Early and his gray-clad troopers, fresh from the looting of York, galloped into Wrightsville, their wagons loaded with captured blankets, a thousand pairs of shoes and much-needed medical supplies. They were bound upriver to join General Ewell's brigade for a frontal attack on Harrisburg.

The aqueduct across the Susquehanna River, connecting Columbia with Wrightsville. *Courtesy National Archives.*

Early—"Old Jube" to his troops—an "unreconstructed Rebel" to the day of his death in 1894, never having taken the oath of allegiance to the United States after the Civil War, eyed the combination wooden bridge and aqueduct connecting the west bank of the river with Columbia.[3] To prevent attack from that direction, he ordered the bridge burned. Pausing only long enough to be sure of its destruction, he gave the order to ride on. The road was clogged with panic-stricken refugees fleeing the Confederate advance, but in the early afternoon he joined forces with Ewell's command.

Burning of the combination bridge and viaduct at Wrightsville–Columbia on the Susquehanna River by Confederate troops on June 28, 1863. From *Leslie's Weekly*, 1863.

They were within three miles of Harrisburg when urgent dispatches were received from General Lee, ordering them to rejoin him immediately. This was the farthest north any Confederate advance ever reached.

In 1864 the Susquehanna and Tidewater Canal enjoyed its most profitable year, earnings amounting to nearly $300,000. This marked the feast before the famine, traffic declining so rapidly that by 1868 there appeared to be no point in continuing what had become a struggle for survival. The Baltimore and Susquehanna, which had become a Philadelphia and Reading line, laid tracks into Harrisburg and captured that business. In 1870 the canal company bowed to the inevitable and leased its property to the Reading for the customary 999 years.

Under railroad management the canal was relegated to the humble role of leisurely coal carrier. As such it produced a small annual profit for many years. Some minor repairs were made, but nothing was done to stay the obvious deterioration of the old waterway. When the severe floods of April, 1894, ripped out its banks, the Reading opened the lock gates for the last time and the Susquehanna and Tidewater passed out of existence. It had had a long and often precarious life. To be exact, fifty-four years had passed since the quavering notes of a captain's horn had warned lock-tenders and their assistants that the first boat was coming down the canal.

Ruins at the mouth of the Susquehanna and Tidewater Canal. *Courtesy National Archives.*

14

The Erie Canal—
Before the Digging Began

OUR SHORT EARLY canals had only a restricted, local significance. Save for the Pennsylvania Main Line, that was also more or less true of those built later. Only the Erie Canal had a national impact—national to the extent of New England, New York, the Middle Atlantic states and the new states of the Middle West. Furthermore, its success supplied the incentive for the construction of every other canal east of the Mississippi after 1825.

Although some half-dozen canals were actually in operation, and as many more under construction, when ground was broken on July 4, 1817, at Rome, New York, the Erie Canal was in every sense a major pioneer undertaking, dwarfing in concept any work of a similar nature previously attempted. Ever since control of that part of New York State lying north of Albany and the Mohawk River had been wrested from the Iroquois, connecting the Hudson with the Great Lakes by water had been a matter of lively debate. At first it was the majority opinion of the proponents of such an undertaking that by way of the Oswego River it should reach Lake Ontario, even though the falls and gorge of the Niagara would block the way to the west; few were bold enough to declare for the longer route to Lake Erie.

It is impossible to say to whom the distinction of having been the first to propose the building of a canal to connect the Hudson with one or the other of the

Entrance to the Erie Canal from the Hudson River. *From Whitford,* History of the Canal System of the State of New York.

lower Great Lakes, Erie or Ontario, properly belongs. There are many claimants. Some sources find the suggestion in the communications of Sir Henry Moore, the royal governor of New York, to his king as early as 1768. Others award the honor to Gouverneur Morris, the soldier and statesman, and cite his remarks made immediately after the Battle of Saratoga in 1777. A better case can be made for Elkanah Watson and Jesse Hawley.

During the Revolution Watson, a young New Yorker, had been dispatched to Paris as an aide to Benjamin Franklin. At the close of the war he remained abroad for several years, in the course of which he made a careful study of French and Dutch canals. Returning home in 1785 he was convinced that a similar system of internal waterways would serve his own country as well as they served Europe. Journeying to Philadelphia, where the future of the United States was being shaped, he was encouraged by the support his ideas regarding inland navigation received from a number of prominent men, including George Washington.

Watson was astute enough to realize that to have any success, he had to settle on a specific project, so thereafter he devoted himself to publicizing the building of a canal across New York State. In 1788 he set out from Albany alone and on foot to explore "the most likely route of such an enterprise." His journal reveals his conception of the needs of the semi-wilderness through which he passed (north of Albany no town had as many as a thousand inhabitants) and "the undoubted blessing better transportation would bestow."

In 1791 Watson published a pamphlet concerning the country through which he had journeyed, calling attention to its needs and how they could be met by the construction of a canal.

He presented copies of his booklet to General Philip Schuyler, recently elected U.S. senator from New York, and Governor George Clinton. They were the two most influential men in the state. When Watson gained their support, the battle was better than half won. Stimulated by such powerful approval, the public not only demanded a canal to the West but a second canal via Lake George and Lake Champlain to Canada.[1]

Responsive to its leaders, the legislature in the spring of 1791 granted charters to the Western Inland Lock Navigation and the Northern Inland Lock Navigation. Although these two companies had separate identities, they really were one, which could not be concealed when Schuyler was elected president of both. The state obligated itself to aid them financially, politics being no different then than today.

The Western Inland Lock Navigation began work at Fort Stanwix and Little

One of the original locks built by the Western Inland Lock Company. *Courtesy New York State Commission for Preservation of Historic Sites.*

Falls in 1793 and soon had five hundred men digging and blasting a channel and installing locks. When it ran out of funds, work stopped until the state came to the rescue by buying $57,500 worth of the company's stock. Work was resumed at Little Falls and was so nearly complete by November, 1795, that boats began to pass through. The canal was less than a mile long, a good half of it cut out of solid rock. In that short distance it had been necessary to install five locks.

In the month remaining before winter brought operations to a halt, the canal collected almost $400 in tolls. Amazingly, the cost of transporting a ton of goods from Fort Stanwix (Utica) to Schenectady dropped from $14 to $5. Immigrants bound for points west also made a substantial saving, because the new Durham-type boats put in use on the canal could move several families and their goods at the same time, something the bateaux, which had to be poled against the current, had not been able to do.[2]

The Northern Inland Lock Navigation made a survey and did some work in the direction of Lake Champlain which was never finished. It had managed to expend a $100,000, all of which was totally lost—due, many said, to the mismanagement of General Schuyler.

The failure of the Lake Champlain project cooled off enthusiasm for building a canal to Lake Erie. Martin Van Buren, the fiery-tongued leader of the Democratic opposition, and his cohorts made political hay out of the debacle. "What reason have the taxpayers of this State to believe that men who can not build a comparatively short canal, through the most favorable country for such an enterprise," they demanded, "be expected to build a canal 350 miles long from the Hudson to Lake Erie, costing millions of dollars?"

Governor Clinton came under attack as one of the sponsors of the canal legislation then on the statute books, but he refused to engage in political mud-slinging and stood by his guns. There was little he could point to with pride: the Champlain enterprise had been a complete failure and the Western Inland Lock Navigation was bogged down in debt. It had done some additional work: a channel had been cut across the portage at Rome between the upper Mohawk and Wood Creek, over which the Iroquois had been toting their canoes for centuries before the first white man had set foot on this continent. In addition to the work done at Rome, a short canal farther down the Mohawk had been completed, and a survey made for a canal around the Cohoes or Great Falls. When that was finished, the Inland Lock would have opened a waterway from the Hudson to the network of lakes and streams in central and western New York and the Genesee Valley. But the company was in such desperate financial straits that building the canal at Great Falls could not be undertaken. In 1810 it petitioned the state for relief. None was forthcoming. Western Lock had some modest accomplishments to its credit, but in seventeen years it had never paid a dividend. Not only had the money received from the sale of stock, state loans and tolls been spent, but the company was also burdened with an indebtedness of nearly $20,000.

Even its friends saw no future for a meandering ditch that was almost as much river as canal. If there was ever to be a waterway across the state to the Great Lakes, it would have to be all canal, following the shortest, most practical route, cutting through or surmounting natural barriers, and using the streams it encountered only as feeders to guarantee the canal an adequate water supply. While this was not a blueprint for the future Erie, it staked out the guide lines.

Not unexpectedly the state purchased the Western Inland Lock Navigation, paying $150,000 for the property, and by an act that barely squeaked through the legislature, was authorized to build its own canal connecting the Hudson with the Great Lakes, making it the first state in the Union to engage in the canal-building business.

But the considerable portion of the citizenry who expected that dirt would soon be flying were in for a long disappointment. A lot of fulsome oratory was expended, and committees were appointed. Little else was done. A grant of $600 was made to Judge James Geddes to make a preliminary survey of a route for the canal. The sum proved insufficient and Geddes had to supplement it with $75 out of his own pocket, for which he was never reimbursed. In his report he advised the Legislature that building a canal to Lake Erie and leaving the Niagara barrier still to be passed would turn the commerce of the West into New York State. On the other hand, if that trade reached Lake Ontario, the bulk of it would continue down the St. Lawrence. It was a sound argument that the Lake Ontario advocates could not deny.

Times were so good it was rumored that President Jefferson was about to ask the Congress to distribute $20,000,000 among the various states, apportioned according to their population. Not waiting for that windfall, Judge Asa Forman and William Kirkpatrick, prominent New York State legislators, were named to visit the president in the hope that building the Grand Canal could be made a joint undertaking of the state and federal governments. Jefferson listened to what they had to say but refused to be swayed in their favor. "It is a splendid project and may be executed a century hence," said he.

> Here on the Potomac, a short canal, sponsored by General Washington, which would be of great good to us has languished for many years because the small sum of $200,000 to complete it can not be obtained. And you talk of making a canal three hundred and fifty miles long through a wilderness! It is little short of madness to think of it at this day! [3]

But the "madness" did not die. Jesse Hawley has been mentioned. As much as any man, he helped to keep it alive. A native of what used to be known as "the North Country"—that part of New York State above Albany—his years of journeying up and down and across it gave him an expert knowledge of its needs and possibilities. A stream of articles extolling the benefits a great canal connecting the Hudson with Lake Erie would bestow flowed from his pen. They appeared in issue after issue of the *Knickerbocker Press,* the *Genesee Messenger* and other newspapers. Although published under a variety of Latin or Greek pseudonyms such as Spartacus or Pro Bono Publico, their style was unmistakable, and the public recognized them as the work of indefatigable Jesse Hawley.

If the commissioners appointed to determine what course the proposed canal should take could not reach an agreement, there was no doubt in Jesse Hawley's mind. Fifteen years later, when the completed Erie was opened to navigation, its course, for most of the way, was the one he had urged.

The Board of Canal Commissioners appointed in 1810 was composed of Stephen Van Rensselaer, Thomas Eddy, Peter B. Porter, Simeon DeWitt, William North, Gouverneur Morris and De Witt Clinton, the last-named then filling the

Governor De Witt Clinton, builder of the Erie Canal. *Courtesy New York Public Library.*

dual role of state senator and mayor of New York City. Clinton's presence on the board was strictly a political payoff. He was already recognized as one of the state's most powerful political leaders. Until now he had been decidedly apathetic about canal legislation, which was more or less reflected in New York City's stubborn refusal to believe that an expensive inland waterway would increase its prosperity. He had been asked to introduce legislation creating the commission and supplying it with funds and he had refused. But he had seconded the measure and with his approval it passed. He was then rewarded by being named to the board and entitled to share in its perquisites.

Clinton was much more than an able politician. At his best he deserves to be ranked a statesman—able to rise above his own personal interests and do that which he believed was best for the people who had elected him, refusing to take the popular course when he believed it to be wrong. His personal integrity was high, as even Martin Van Buren, his deadliest political foe, acknowledged. The many writers who have told the story of the Erie Canal and its builder have portrayed him as a relentless fighter who refused to acknowledge defeat, no matter how great the odds against him. And the odds were often tremendous.

Leadership came easily to De Witt Clinton. Born at Little Britain in Orange County, New York, he found his first employment on graduating from Columbia as secretary to his uncle, Governor George Clinton. In 1798 he was elected to the New York State Assembly and the following year he advanced to the state Senate. He won reelection to the Senate and in 1803 began his remarkable reign as five-time mayor of New York City, following which he was twice elected governor of New York.

De Witt Clinton was forty-one when appointed to the Canal Commission. The negative or at best lukewarm position he had taken in the past on the building of internal waterways convinced supporters of the proposed canal that they could not expect any help from him. But as the commissioners explored the country through which a canal might pass and their findings were published, the canal faction was amazed to discover that Clinton had become their staunchest spokesman.

This startling about-face on his part has never been satisfactorily explained. It would seem to be a reasonable conjecture, however, that the vast realm of fertile, timbered land stretching away to the horizon in all directions, which he was seeing for the first time, and which obviously was waiting only for the axe and the plow to be made productive, so impressed him that he believed to open it to settlement by providing a cheap means of transportation was the paramount duty of the state.

As mayor of New York City he was in an excellent position to arouse support for such an undertaking.[4] Men who had become prominent in the cause of the so-called Canal Party, recognizing his ability as a leader, capitulated in his favor, and at the mass meetings that were held that autumn he was the dominant figure. Realizing that the great obstacle to success was regional jealousy, Clinton assured the northeastern counties that a Lake Champlain Canal (the old Northern Lock Navigation) would be built. To win over the southern-tier counties along the New York-Pennsylvania boundary, who saw no benefit to themselves in a mainline canal to Lake Erie, he promised that they would be incorporated into the system by branch canals.

In all his public appearances he urged his audiences to address petitions to the legislature demanding that construction of the Grand Western Canal be undertaken forthwith and that the Board of Canal Commissioners be authorized to borrow money for its construction. When the legislature met in January it was confronted by a mountain of mail. It was so preponderantly in favor of building the canal that there was little doubt of the outcome. A bill authorizing the Canal Commission to begin construction and pledging the credit of the state to the extent of $5,000,000 passed both houses by a comfortable margin. The disgruntled opposition at once labeled the undertaking "Clinton's Ditch." Van Buren denounced it as "the Ditch of Iniquity."

But the battle was not yet won. There were delays, some necessary. Before work actually began, the War of 1812 intervened. The British had been stopping our merchantmen, impressing our seamen and commandeering our ships. The provocations became unendurable. Warnings having failed, the Congress declared war on Great Britain in the summer of 1812.

This so-called second war for independence ended all talk about internal improvements for the time being. New York State, exposed to invasion from Canada, felt the war's impact more than most states. American troops marched into Canada and won minor victories at Lundy's Lane, Queenstown and Chippewa. But it was not until after Commodore Perry's victory on Lake Erie and Macdonough's annihilation of the British fleet on Lake Champlain that the American frontier along the Niagara River could be considered safe.

In February, 1814, taking advantage of the stringencies caused by the war, opponents of the canal succeeded in revoking the authority given the Canal Commissioners to borrow money for construction purposes. It was a serious setback, for without money little could be done. Despairing of accomplishing anything, the canal commissioners resigned. Clinton refused to accept defeat. At his invitation, several hundred prominent men attended a meeting in New York City early in December, 1815, to discuss ways and means of reopening the canal question. The war had been over for a year but business had not revived as expected. In his remarks Clinton declared that building the Erie would supply business with the spark it sorely needed.

Interest in the canal, dormant for four years, suddenly revived. Again canal mass meetings were held in villages and towns throughout the state. In April, 1816, the legislature passed an "Act for improving the Internal Navigation of this State," under the provisions of which a board of five commissioners (of whom Clinton was one) was "empowered to scheme a communication by canal between the Hudson and Lakes Erie and Champlain." The commissioners were given $20,000 for surveying expenses. They were *not* given authority to raise other moneys or to begin actual construction. But the door to that end had been opened.

It was a victory for Clinton, not won on the merits of the canal question but rather by his shrewd political horse trading. Governor Daniel Tompkins had been nominated for the vice-presidency of the United States as James Monroe's running mate. To win election, the ticket needed the votes of New York State, and that meant the active support of Clinton. To gain this support, Tompkins dropped his opposition to the Canal Act, and indeed came out in its favor. Seemingly the arrangement did not end there, for Tompkins was no sooner elected vice-president than he resigned the governorship, and the following spring, in a special election, De Witt Clinton was swept into office as his successor.[5] Victory was the sweeter because it had been won over Martin Van Buren who had waged a bitter campaign to defeat him.

Van Buren knew that without any loss of time Clinton would ask the legislature to amend the Act for Internal Navigation then on the books to give the commissioners authority to begin construction at once, allocating $5,000,000 to cover the cost of the project. In essence these provisions had been made in the Act of 1812, which had been rescinded. When Clinton appeared before a joint session and made his proposal in person—informing the legislators that the Holland Land Company, owners of half a million acres, had offered to give the canal a free right-of-way a mile wide through its immense holdings—Van Buren realized that for opponents of the Erie this was the last stand. He believed he had the votes to defeat the measure but he was not sure; it was going to be very close.

For a month, with the so-called Albany Regency, Van Buren's and Tammany Hall's powerful political machine, leading the fight to kill the pending legislation, a bitter, acrimonious debate dragged on. On the day before adjournment, with no decision having been reached, Van Buren got to his feet, and placing the public interest above political partisanship, stunned his followers by casting his vote in favor of the canal bill. It broke the logjam and launched New York State on what was to be its greatest undertaking—the building of the Erie Canal.

It was a victory for De Witt Clinton. It was also a victory for Martin Van Buren.

In the Mohawk Valley—open door to the West. *Courtesy New York Public Library Picture Collection.*

15

The Building of the Erie

THE RANCORS AND animosities left by the stormy passage of the Canal Act manifested themselves at once. The Board of Commissioners was reduced to five, Clinton and four others,[1] and as a check on their spending, a new authority, Commissioners of the Canal Fund, was created.

Since the new agency was to have its fingers on the purse strings, it was freely predicted that the arrangement would result in endless bickering and unnecessary delays. The public, interested only in having the canal completed as quickly as possible, resented this bit of political harassment, the feeling being widespread throughout the state that if the Erie Canal—or Grand Western Canal, as many preferred to call it—was to be built, tall, commanding De Witt Clinton was the man for the job and should be assisted, not impeded.

In the early spring of 1817, facing the greatest undertaking yet attempted in America, and about which he had neither previous experience nor knowledge, Clinton exhibited remarkable self-confidence. To the nonbelievers—not necessarily unfriendly to him personally—he declared that building the Erie was a dream that would be realized, and for the more numerous skeptics who were of the opinion that "the canal may be completed some day but not in our lifetime," he had only one answer: "What has been accomplished elsewhere we shall accomplish here; the day will come in less than ten years when we will see Erie water flowing into the Hudson."

It had been assumed by many, including the managers of the canal fund, that the improvements made by the defunct Western Inland Lock Navigation, now the property of the state, would be incorporated into the Erie Canal. Clinton nipped that idea in the bud by announcing that "we do not propose to burden the Erie with the blunders made by Western Inland Lock. However, if it should be found expeditious to make use of certain portions of the old works, we will; but we shall be under no compulsion to do so."

This was the initial confrontation between the two authorities. There were to be others. Such squabblings, politically motivated, were minor annoyances compared to the almost daily diet of crisis after crisis that the builders of the Erie were to face.

Judge Benjamin Wright was appointed chief engineer and James Geddes, a lawyer, named assistant chief. Geddes was put in charge of work on the Champlain Canal and Wright was assigned to the Erie. They drew about them a number of eager young men, among whom were John Sullivan, John Jervis, Frederick Mills, Nathan Roberts and Canvass White. In the years to come all were to garner a measure of fame as graduates of the so-called Erie School of Engineering. When they signed on they knew how to squint through a theodolite, but were in no sense engineers. For that matter, neither were Wright and Geddes. Both were competent surveyors, self-taught and ambitious. Profiting by experience, in a few years they were recognized as two of the country's foremost hydraulic engineers.

Wright completed the final survey of the Erie that spring and established the official length of the canal from Buffalo to what was to become the canal basin at Albany, by means of which boats would enter the Hudson River, at 363 miles; the descent from Lake Erie was measured at 555 feet. Two-way traffic—that is, from west to east and east to west—was to be achieved by installing eighty-three locks; twenty-seven of them in the first fifteen miles between Albany and Schenectady around the Cohoes Falls.

Although Wright's figures showed a slight refinement of those previously submitted by him, they were essentially the same, and the canal commissioners accepted and published them, together with a map showing the course the Erie was to take. From Albany it was to point northward a few miles to Troy, then strike westward up the Mohawk Valley for Schenectady, Amsterdam, Canajoharie, Little Falls, Herkimer, Utica, Rome, Oneida, Syracuse, the Montezuma marshes, Lyons, Rochester, Lockport and Buffalo—place names with which all of us are familiar. Today these cities and towns dominate the heartland of central New York State, but in 1817 they were only villages. The whole region was so thinly settled that there wasn't a stage line west of the Mohawk; only Canandaigua, at the head of Seneca Lake, and Batavia showed some evidence of developing into prosperous small towns. On Lake Erie the great port of Buffalo was still in its infancy; its population combined with neighboring rival Black Rock amounted to fewer than seven hundred.

Employment and purchasing offices were opened in New York City and Albany. The purchasing of horses, mules, wagons and tools began. In New York City in May Samuel Young, treasurer of the Canal Commission, awarded contracts for excavating the channel. The contracts were given under bond as a guarantee that the work would be completed in a satisfactory manner within a

Ruins of the aqueduct at Little Falls, N.Y. *From Whitford,* History of the Canal System of the State of New York.

stipulated time. A prime contractor having engaged to build ten miles of channel, berm and towpath could, and usually did, parcel out the work to subcontractors who were responsible to him and not to the Canal Commission. The price paid per yard for digging varied according to the terrain and the obstacles to be encountered. It did not include building locks, bridges and aqueducts, which the state was obligated to do.

We never tire of hearing how the Irish bogtrotters built the Erie. Certainly a great number of them—perhaps more than three thousand—put their sweat, blood and muscle into it. As they fought their way through the mosquito- and malaria-infested Montezuma marshes west of Syracuse, toiling in waist-deep muck and water, wearing only a shirt and slouch cap to shield them from the relentless sun, they wrote a page of human endeavor that has seldom been equaled. And for this they were rewarded with the princely wage of $8 a month—or to be more exact, for twenty-eight rainless days of work—and the privilege of sleeping on the floor of a $15-shack along with a dozen others of their kind, their food the cheapest and coarsest the contractor could provide. As a bonus a tot of whisky was doled out to them every two hours—to keep them going.

No one has thought to give Tammany Hall, New York City's powerful Irish political club, credit for the part it played in the building of the Erie. In the first quarter of the nineteenth century successive failures of the potato crop had spread famine throughout the western counties of Ireland. To escape the threat of starvation thousands of young Irishmen emigrated to America. Being penniless, they had to depend on relatives living here for their passage money. Thousands who had no such relatives were brought over by the rich Irish contractors of Tammany Hall, which meant that they arrived in the United States as indentured men, obligated to repay by their labor the money that had been advanced to them. It was largely from their ranks that the bogtrotters of the Erie were recruited and shipped up the Hudson to Albany.

There were two long levels on the Erie—stretches of the canal not requiring locks. The longest ran from Frankfort (just west of Herkimer) to Syracuse, a distance of 69½ miles, always referred to by "canawlers" as the Long Level.[2] The next longest, sixty-two miles, extended westward from Rochester. Construction could have begun at a number of places. Clinton wisely decided that it should begin at Rome, on the Long Level where the digging was easiest and the greatest progress could be made in the shortest time. By the end of the year, he was determined to have something to show the voters of New York State.

On July 4, 1817, with ceremonies appropriate to the occasion, Joshua Hathaway, president of the village of Rome, handed a spade to Commissioner Samuel Young, secretary of the Canal Commission, who spoke at length before passing the spade to Judge Richardson, the canal's first contractor. The judge plunged the blade into the sod and lifted the first shovelful of Erie earth. By December fifteen miles of canal had been completed. It was an excellent beginning, but Clinton felt the need for even heavier ammunition to fire at his opponents in the upcoming session of the legislature, now only several weeks away. To that end he called his chief lieutenants together at Albany and for several days discussed plans for the resumption of work the following spring.

In late summer he had divided the Erie into three divisions and placed a key man in charge of each. Geddes had been transferred from the Champlain job and assigned to the western section (Lake Erie to the Seneca River) and Wright made responsible for the middle section (Seneca River to Rome). The eastern section (Rome to Albany) had been placed under the supervision of Charles C. Broadhead. Colonel G. Lewis Garin had succeded Geddes on the Champlain Canal and William Peacock sent out to Buffalo, not only to locate and build the western terminus of the canal but also to discover the most feasible route by which the Erie could climb up and down (or around or through) the rocky, eleven-hundred-foot Niagara Escarpment that stood squarely in its path.

Other Erie luminaries were present at the Albany conference, among them Nathan S. Roberts who was to achieve enduring fame as the creator and builder of the famous Lockport combines—five pairs of double locks cut through solid rock by which ascending and descending boats, passing side by side, conquered the Niagara Escarpment. Youthful Canvass White (he was twenty-six at the time), the genius of the Erie engineers, was not present, Clinton having sent him to England the past summer to study English canals and especially lock construction and operation.

Clinton's main purpose in calling the conference was to lay before it his plan for opening work in two separate locations the following year: Wright to work east from Syracuse (still called South Salina by many) and Broadhead to work westward to meet Wright's crew. By that arrangement Clinton believed it would be possible to complete the 69½ miles on the Long Level by the time winter brought operations to a halt. Being politically oriented, he was aware of the impact such

Benjamin Wright, chief engineer of the Erie and other canals. *Courtesy Library of Congress.*

James Geddes, one of the famous triumvirate of Erie engineers. *Courtesy Library of Congress.*

An early photograph of the Lockport "fives."

Combined locks on the Erie Canal at Lockport. *Courtesy New York Public Library Picture Collection.*

Nathan S. Roberts, whose greatest engineering achievement was the building of the Lockport "fives." *Courtesy Library of Congress.*

a program would have if it could be accomplished. He was also aware how disastrous it would be to announce such a goal and then fail to achieve it. He glanced around the table and addressed himself to Wright.

"I can't guarantee the weather," Wright told him. "But if we get another good April like we did this year, we should be able to complete the job before snowfall. We can build more stump pullers during the off months, but we'll have to look to you to supply the hundreds of dump wheelbarrows and small tools we'll need."

"You build your stump pullers," Clinton told him. "I'll put the tools in your hands." [3]

When he announced that the Erie would be completed from Syracuse to Herkimer that year, his political enemies thought they had him out on the end of a limb. Even his well-wishers doubted that he could make good. But he did; the Erie Canal was becoming a reality.

As it became increasingly apparent that the Erie was going to be pushed through to completion, hundreds of Yorkers left their farms in the care of their wives and growing sons and sought employment on the canal. They brought that priceless quality which, for want of a better name, has been described as Yankee ingenuity. Pioneers or the sons of pioneers, schooled to overcome any difficulty that confronted them, they invented the stump-puller, an ingenious device that enabled a crew of half a dozen men and a team of horses to pull and remove thirty to forty stumps per day.[4] Hardly less important as a time-saver, they demonstrated that by attaching a cable to the top of a sixty-foot tree and winding it up on an endless screw a man could fell it single-handed. Instead of wearing out a team bucking brush, they made the going easy by adding a horizontal cutting bar to the plowshare. Wright and his staff shook their heads and wondered why they hadn't thought of it.

As in canal construction elsewhere, the only explosive used on the Erie was Dupont's Blasting Powder. It was an improvement over black powder but a far cry from dynamite and nitroglycerin, which were not yet available. The Irish did most of the blasting. They enjoyed the dangerous work and were recognized as the best "blowers" on the canal. When an accident occurred it was usually traceable to recklessness.

Acting on Peacock's report, Governor Clinton appointed a commission to survey and plan a harbor at Buffalo. The decision brought howls of protest from neighboring Black Rock, which called up its company of militia and threatened to bar the passage of the commissioners when they appeared. But cooler heads prevailed. As Buffalo boomed, Black Rock settled back as a suburb of its more fortunate neighbor.[5]

When Canvass White returned from England in 1818 he brought with him new instruments, a sheaf of carefully executed drawings and a better knowledge of canal construction—especially locks—than any other man in America possessed. He found Clinton and his subordinates in a quandary over whether to build the Erie locks of wood or stone. Hydraulic cement, necessary if stone was used, could be procured only from Europe and at great cost. Wood, on the other hand, was so perishable that it could not be expected to last more than a few years. As a compromise the decision was made to build the locks of stone, putting the blocks together with common mortar and merely "pointing" the joints with costly imported hydraulic cement.

White was placed in the unhappy position of knowing it was a bad decision, that if soluble mortar was used the blocks would quickly deteriorate and fall apart, but feeling that if he spoke out he would be put down as a young upstart bent on promoting himself. He also knew that William Weston, the English engineer, had found a deposit of trass, the principal ingredient of hydraulic cement, in Massachusetts when he was building the Middlesex Canal. Based on what little he knew about geology, White thought it possible that trass (volcanic pumice) might be found in New York State.

In the time he could take off from running levels and his other duties, White began scouring the countryside for the mineral. He had no luck and was about ready to give up searching when he fell in with two young men of the village of Chittenango who described to him a gray substance they had seen. Excited by their disclosures, White was directed to the location of the deposit. A hasty examination convinced him that it was volcanic pumice. When a small quantity submerged in water overnight turned into solid rock, any doubt of what he had found vanished.

History does not reveal the identity of the young men who should have shared in the acclaim that came to Canvass White for the Chittenango find. Its importance to the Erie cannot be overstated, for the excellence of the canal's stonework would not have been possible without it. The proximity of the discovery to the main channel of the Erie was an additional stroke of good fortune, full advantage of which was taken by the construction of the short Chittenango Canal.

Compared with other canals already in operation or under construction, the dimensions of the Erie were generous. It could not have occurred to anyone in those early years that the time would come—and soon—when they would prove inadequate and would have to be enlarged—not only once but twice, even three times if you include the modern New York State Barge Canal. The original figures called for a channel forty feet wide at top, twenty-eight feet at bottom, with a depth of four feet. The width of the towpath was fixed at ten feet.

In central New York State there were so many creeks and small streams, most of them flowing from west to east, that could easily be converted into feeders for the Erie that there was never any question about it having an adequate water supply. In the years before it became an important branch canal in its own right, the twenty-two-mile stretch of Black River north of Rome was used as an important feeder.

The season of 1819 got off to a discouraging start. It began with seven days of rain that caused many streams to overflow their banks. In places the freshets spilled over into the canal bed and damaged the channel. West of Syracuse several thousand men had been gathered for a grand assault on the Montezuma marshes but conditions were so bad that work could not begin. It was the middle of May before sufficient water had seeped away to permit the bogtrotters to move in. The muck they dug up had to be removed and heavy earth brought in to build up the banks.

As June faded into July the weather turned hot and humid. A miasma hung over the marshes. Whisky could not keep the workers going; they dropped their tools and dragged themselves to their barracks. They were prostrated by malaria, ague and typhoid fever. Very little medical assistance was available. Over a thousand men were stricken before the pestilence ran its course. Many died.

Several months passed before work could be resumed in the marshlands.

Elsewhere it was prosecuted with vigor. To dispel the feeling that the Erie had suffered a serious setback, Clinton resorted to a bit of showmanship. Under his direction the first boat to float on Erie water was built at canalside at Rome. Sixty feet long, it was to serve as a model for hundreds that were to follow. Gaily painted and named *Chief Engineer of Rome,* in honor of Benjamin Wright, it stood ready to be launched when the Black River feeder (later to become part of the Black River Canal) was opened and water let into the completed stretch of the Erie between Rome and Utica on October 22.

The *Chief Engineer*'s passengers included Governor Clinton and the commissioners—"attended by many respectable gentlemen and ladies." The *Chief Engineer,* drawn by a single horse, reached Utica without incident and returned to Rome the following day. Wrote a correspondent of the *Genesee Messenger:* "The scene was extremely interesting and highly grateful. The embarkation took place amid the ringing of bells, the roaring of cannon and the loud acclamations of thousands of exhilarated spectators, male and female, who lined the banks of the newly created river. The scene was truly sublime."

In more practical terms, it accomplished Clinton's purpose. Anxious to strike a second time while the iron was hot, he announced that the Champlain Canal would be open to navigation from Watervliet to Whitehall, its entire length, by way of Wood Creek, Fort Miller Falls and Saratoga Falls, sixty miles in all, by the end of the year. He was gambling on the weather, for actually considerable work remained to be done and could be accomplished before January 1 only if the mild temperatures of the past month continued for another two or three weeks. Luck favored him, and several boats did pass through the new canal in early December.

Completion of the Champlain Canal in so short a time was hailed by pro-canal editors as a great achievement. Opposition newspapers could not ignore the fact but they made much more of the news that the Holland Land Company had

The Champlain Canal at Whitehall, New York, about 1840.

granted 100,632 acres of land west of the Seneca River "in aid of the Erie Canal." It was alleged that this and similar grants were not being made in the public interest but for private gain and that the course of the canal was being determined by secret understandings between the canal commissioners and the land speculators. No proof of this was ever offered.

Although cutting the channel through the Montezuma marshes had begun so disastrously, enough work had been accomplished to convince Wright and his staff that it could be completed the following year. On that assurance contracts were let for construction of the middle section of the Seneca River and the opening of the Salina side-cut.

It had been quite a year after all for Clinton and the Erie.

Junction of the Erie and Champlain canals. *From Whitford,* History of the Canal System of the State of New York.

16

The Great Enterprise
Goes Forward

ALBANY WAS THE main gateway to central and western New York and the sparsely populated states beyond its borders that had been carved out of the old Northwest Territory. It therefore could not escape being aware of the increasing Yankee migration from New England nor of the arrival of thousands of foreign-born immigrants, Dutch, German, Scottish and Scandinavian, who were being brought in by the land companies and colonized so that they would have neighbors who spoke their language and were of the same religious persuasion as themselves.

Observers found little reason to doubt that these men and women who had had the courage to cross the sea in hope of finding a better life than they had known in their homeland would be quickly assimilated and make good, thrifty citizens. It was different with the lean, toil-hardened ex-Massachusetts and Connecticut Yankees who were pointing their wagons up the Mohawk road every day. Their minds were set on the Ohio country and the virgin lands beyond. Not one in a hundred had a definite idea of whither he was bound and how he was to get there. That many of them would end their westering and put their roots down within the borders of New York State seemed unlikely.

But many did. The promise of the future prosperity that was to come with the completion of the Grand Canal had something to do with it. Already old villages were growing and new ones being established, the Erie being re-

Lockport, Erie Canal. *Courtesy New York Public Library Picture Collection.*

sponsible for their nautical names—Lockport, Gasport, Brockport, Fairport, Port Byron, Middleport, Forestport and many others.

Although to label a man the "father" of this or that for some singular achievement is to resort to a cliché and new words have been coined to replace it, none conveys the respectful recognition of the original. Certainly De Witt Clinton was the father of the Erie. If not for him the great waterway would not have been built. It would have been debated without anything being done until the railroads took over and there was no longer any need for a canal.

Clinton's popularity irked envious politicians in his own party as well as in the opposition. His enemies, personal and political, formed a coalition that gave them control of the legislature. They removed his supporters from state office and supplanted them with men who opposed him. They appointed a commission to report on the advisability of ending the Erie Canal at Rochester, the already flourishing "Flour City," on the Genesee and connecting it with Lake Ontario by that river.

Rochester, not wanting to be cut off from the West, rose up in arms. Syracuse, sitting on top of one of the greatest salt deposits in the United States, did likewise. Destined to become two of the most important industrial cities in the state, they already had their eyes fixed on the world markets the Erie Canal would make possible. The furor raised against this latest attempt to cripple the canal sent the proponents of the measure scurrying to cover and no more was heard about it.

Early in 1820 the Erie was opened to navigation from Utica to Rochester. Rochester boatyards had been busily building boats in anticipation of the event. The first packet to navigate the canal, the "elegant" *Lion of the West* (seventy-six feet in length with a beam of fourteen feet), left the town for which it was named on the morning of April 21 and with only one change of horses reached Lyons in late afternoon. After laying over for the night, it arrived in Syracuse the following day, from where it proceeded to Utica, maintaining an average speed of five miles per hour.

Even that modest rate of progress created a wash that ate into the freshly dug banks. The canal commissioners therefore set the legal speed limit at four miles an hour and established penalties for exceeding it when they issued their *Rules and Regulations for Navigating the Erie Canal*. A schedule of tolls was published and toll collectors appointed to receive them. No weigh-locks having been installed as yet, tolls had to be levied on the value of the cargo a boat was carrying rather than by weight.

It was reported that seventy-three boats took part in the great Fourth of July celebration at Syracuse in 1820, "many delegations from other countries arriving by water." If so, most of them must have been very small, probably bull-boats (so named because of their rounded bow), which could be built for a few hundred dollars.[1] But putting the exaggerations aside, there is bountiful evidence that the Erie was being put to use as soon as the various sections were thrown open to navigation, as was noted by a Utica newspaper:

> Our village on Friday, twenty-fifth inst., presented a scene of bustle and stir never before witnessed here. On Saturday the packet boat from Rochester left here with 84 passengers aboard on her first trip. A boat will leave this place every morning, Sundays excepted, during the season and continue through to the Genesee River.

The Erie Canal at Rochester. *From Whitford,* History of the Canal System of the State of New York.

Clinton had been present at the Syracuse celebration, basking in the acclaim that was rightfully his. Because he was overconfident or ignorant of the cabal that was forming against him, he declared in his remarks to the assemblage that the Erie Canal would be completed in 1823. It was a political blunder, placing him in a position from which he could not retreat, and his enemies lost no time in letting him know it.

Undoubtedly there was somee substance to the allegation that Clinton had become so involved in building the canal that he was neglecting his duties as governor. Nor could it be denied that while he had completed some sections of the Erie he had put off coming to grips with its two knottiest problems—the Niagara Escarpment and the wild falls of the Mohawk.

As the attacks on him became more virulent, Clinton countered by announcing that the fifty-one miles of the western division were half-completed. A few weeks later he was able to state that water had been turned into nine miles of the new channel. He was running to succeed himself as governor, but being confident of winning, had been doing very little campaigning. The opposition had induced the former governor, Daniel Tompkins, who was still vice-president of the United States, to run against him. Under the circumstances Tompkins' candidacy was suspect. Clinton did not take it seriously. But on October, his political lieutenants informed him that he was in serious trouble, that Tompkins was making a race of it and was far ahead in New York City.

With election only a few weeks away, nothing much could be done. Clinton wisely decided to rise or fall on the record of what he had accomplished on the Erie, and to tilt the scales in his favor, he ordered work begun on the troublesome fifteen miles of construction between Schenectady and Albany—fifteen miles

Weighlock Building, Syracuse, 1850.

by stagecoach but considerably more by the route the canal had to take with its two viaducts across the Mohawk and down twenty-seven locks. "It's ridiculous." the Tompkins camp jeered. "A man could walk from Albany to Schenectady and get there ahead of one of Clinton's scows."

Upstate New York suddenly realized that it was the canal as well as Clinton personally that was under attack. That decided the issue and Clinton was re-elected by a scant plurality of several thousand votes.

In what must be considered a reflection of Clinton's victory, the legislature authorized the canal commissioners to borrow $1,000,000 from the state in 1821 and another $1,000,000 in 1822. With abundant funds available, construction of the Erie proceeded at a quickened pace. By midsummer its work force exceeded thirty-five hundred men. The section of the canal between Utica and Little Falls was opened to navigation. This was followed several months later by extension of the channel to Schenectady.

Although by the end of the year some feeders had been dug, dams had been built to control the water and the aqueduct at Little Falls had been completed, a vast amount of work remained to be done. However, hundreds of boats were plying the long stretch of the Erie between Rochester and Schenectady. Noble E. Whitford, the best authority we have, places the number at fourteen hundred. They came down the canal laden with flour, salt, smoked hams, bacon and other farm produce. At Schenectady the boaters found a clamorous throng of settlers, native and foreign-born, waiting to bargain for transportation west.

For years there had been a prosperous wagon trade over the Mohawk Pike between Schenectady and Albany. Under the impetus of the canal trade it suddenly became a booming business.

Before winter brought operations to a halt for 1821, the Erie Canal had been opened for 220 miles, and work had begun on the great viaduct over the Genesee River and on the eleven-hundred-foot-long viaduct that carried the canal back across the Mohawk at Cohoes. Hundreds of bridges had been built. Great as these strides forward were, so much remained to be done that it began to appear

Aqueduct on "improved" Erie Canal at Little Falls, N.Y.

The great Erie Canal viaduct across the Genesee River at Rochester.

doubtful that Clinton could make good his pledge that the waterway would be finished in 1823.

Much of the folklore of the Erie is woven around the multiplicity of bridges built by the canal commissioners. Eighty were constructed between Little Falls and Schenectady alone. In the vernacular of the day they were referred to as "occupation" bridges, that is, they were constructed by the state in fulfillment of a pledge to provide farmers whose lands had been severed by the canal with bridges for cattle crossing and pedestrian traffic. But the economy-minded state did not feel obliged to build them very high.[2] They were so low, in fact, that a person standing on the deck of a canal boat had to stoop in passing beneath one or risk being decapitated. Packet-boat passengers quickly learned to heed the captain's bellowing cry: "Low Bridge! Everybody down!"

The famous Lockport "Fives" (or "Combines" as canawlers preferred to call them) are generally regarded as the single most spectacular accomplishment of the Erie engineers. Hardly less impressive, however, was the feat of taking the canal down to the Hudson from Schenectady. That Benjamin Wright and his staff found it necessary to install twenty-seven locks to reach Albany, only fifteen miles away by the crowded overland road, speaks for itself and needs no embellishment. The curving course of the Mohawk was a further impediment. At Vischer's Ferry, where it began a northerly looping around the town of Cohoes, it had to be recrossed. The only way that could be done was by a viaduct.

Canal historians have described the beautiful Genesee River Aqueduct, with its eleven Roman arches (originally only nine), but have had very little to say about the longer but somewhat less elegant aqueduct at Cohoes. It was built on stone piers that carried canal traffic across the Mohawk, twenty-five feet above the river's high-water level. Of the three Erie aqueducts, it was the longest: 1,188 feet; the Genesee measured 802 feet and the Little Falls Aqueduct 744 feet.

A *crossing,* as the word implies, occurred whenever the canal encountered a stream of flowing water cutting across its path on the way to join the Mohawk. Although such interruptions were numerous, few were a serious problem. If they could not be converted into valuable feeders, they were bypassed by short, inexpensive viaducts. There was one, however, the feared and notorious Schoharie Crossing, that was not to be taken lightly. The accidents and tragedies that occurred there, both before and after the building of the Schoharie Dam, were woven into the very fabric of Erie history, fiction and balladry.

Schoharie Creek appears innocent enough as it meanders down the beautiful Schoharie Valley to its conjunction with the Mohawk in the vicinity of old Fort Hunter. But its drainage basin is so great that several days of hard rain will send it over its banks even today. In 1821, soon after the Erie came through, a bad freshet sent Schoharie Creek on a rampage. A boat and its horses were washed away. The crew escaped with their lives but the team perished. A week passed before the boats could proceed. By then upwards of fifty had been caught in the Erie's first traffic jam. Fights broke out as captains and crews strove to gain an advantage over rival boats.

A "modern" canal boat, 260 feet long. *Courtesy National Archives.*

To contain Schoharie Creek in the future, it was decided to lock it up behind a dam. Canvass White was entrusted with the construction. Surveys were made and some work done, but the dam was not completed until the following year. Based on piles and built of timber and stone, it was 650 feet long—a costly improvement that was only justified when the canal bed was raised and excess Schoharie water led into the river by tunnel.[3]

Perhaps no one realized sooner than Clinton himself that he was not going to be able to fulfill the promise he had made so confidently at Syracuse on the Fourth of July in 1820, that the Grand Western Canal would be completed in 1823. He still had twelve months to go before being called to account, but he chose not to wait. Addressing the opening session of the legislature, he surprised his opponents by acknowledging that the Syracuse promise could not be met. But measuring the amount of work that remained to be done against what had been accomplished, he declared that the Erie Canal could and would be completed in 1825.

His enemies, believing they had found a chink in his armor, used it to tarnish his image. Over $6,000,000 (a vast sum for those times) had been expended on the Erie Canal already, they protested, and the end was not yet in sight. What reason was there for believing that Clinton would not be back in 1825 asking for more time and money?

The most effective counterargument was that having gone this far it would be folly to draw back now. That opinion prevailed and reluctantly $1,300,000 was appropriated for the canal fund.

When the Genesee Aqueduct was completed, early in 1823, the sixteen miles of channel that had been dug west of the river became another link in the navigable length of the Erie, water being let in by opening the Genesee headlock. The extension was used at once for bringing men and supplies up to the scene of the continuing digging, which was pointing westward for what were to become the prosperous canal-made towns of Brockport and Lockport.

Out at Buffalo work continued on improving the harbor and providing a suitable terminus for the Erie. The mouth of Black Rock Creek was widened and wharves constructed to expedite the transferring of freight from lake steamers and sailing vessels to canal boats. As more and more dockage continued to be built, wide doubt was expressed that the need for it would ever be realized. The Grand Western Canal was in full operation for only two years when the improvements made at Buffalo were found to be grossly inadequate.

In August, 1823, six years and a month after the first sod had been turned at Rome, ground was broken at Buffalo and the digging of the western section of the canal begun. Across the state, at the eastern end of the great waterway, construction had reached the Hudson, eight miles north of Albany. At Albany, in cooperation with the capital, work was begun on a commodious canal basin.[4]

Today, in that highly industrialized region between Cohoes and Albany, the visitor is often at a loss to tell when he is out of one town and into the next. That condition did not obtain in 1823, but even then its natural advantages were apparent. The Mohawk and the Hudson had their confluence just above Troy, and across the river at what is Watervliet today (then known as West Troy or the Arsenal City because the U.S. Army maintained an arsenal there) the Erie and Champlain canals formed their junction. Evidence enough, it would seem, that the commerce of the state must flow that way.

That autumn a memorable event in the history of the Erie and Champlain canals occurred. On the morning of October 8, the surrounding countryside arrayed in its gaudiest raiment, and all arrangements for a monster celebration having been made, a flotilla of gaily decorated canal boats from the north and west passed down the junction canal into the still unfinished basin at Albany.

They were saluted by cannon fire, martial music and the dignitaries of the state and town. That evening there was a grand ball, fireworks and a banquet. Clinton was very much in evidence. Even his enemies admitted privately that he had done well. But they were not done with him.

In the few weeks that remained before the winter tie-up occurred, the capacity of the junction canal was tested. More than two thousand boats are said to have arrived and cleared from the Albany "long dock." Tolls collected by the abbreviated Erie that year amounted to $36,000. Obviously they were bound to increase, but had anyone been rash enough to predict that they would top $750,000 in another three years, he would have been dismissed as a fool. Incredibly, they did; the combined tolls of the Champlain and Erie canals totaled $765,000 in 1826.

Long before the Erie built into Albany, Schenectady had become recognized as the unofficial eastern terminus of the canal's freight and passenger traffic. Understandably the town, concerned about its continuing prosperity, viewed with misgiving the extension of the waterway to the Hudson. As feared, Schenectady lost most of the freight business, but practically none of the more lucrative passenger trade. By stagecoach the fifteen-mile journey from Albany could be made in three-quarters of an hour; via the canal, with twenty-seven locks to be negotiated, the passage was long and tedious. It became routine for travelers to leave the capital in the late afternoon, put up at the Givens Hotel in Schenectady (by all reports an outstandingly good tavern, famous for the meals it set), spend the night there and board a packet in the morning for the journey west.

Remains of Montezuma Aqueduct on the original Erie Canal. *Courtesy New York Public Library Picture Collection.*

Competition among the various captains for passengers became so keen that many employed "runners," usually brash young men, to steer travelers to their respective boats. In addition to the gentry going by packet, there were the emigrants, with their families and belongings bound for the West, who had reached Schenectady by wagon. "It was not unusual," we are told, "to see hundreds of groups gathered on the bank, guarding their goods and bargaining with one captain or another for a concession of his rates. If it was a group of Dutch, Swedes or Germans who spoke no English, there was usually a representative of the land company that had them in tow on hand to speak for them." [5]

The average charge was one and a half cents per mile for adults. Children under five were carried free. Adults were allowed a hundred pounds of baggage, above which there was an additional charge. The latter was flexible enough to permit some bargaining.

Up to 1824 navigation of the Erie and the Champlain Canal was limited to the daylight hours. On the Champlain, which was comparatively short (sixty miles), the freighter was the only type of boat in operation. Freighters carried deck passengers but had no accommodations for sleeping or feeding them. On the Erie, three types of boats were in operation: the fast, exclusively passenger-carrying packets; the long-haul freighters; and the small, individually owned short-haul freight boats that made a business of picking up and delivering mixed cargoes whatever their destination. The fast-traveling "line" boats, with their relay stations where fresh horses and crews were always waiting, and which were to lord it over the Erie eventually, had not yet appeared.

In their demand for greater speed on the canal, the boat lines, of which there were many, forced the commissioners to keep the locks open night and day. Their fast-moving boats damaged the berm and towpath so frequently that a limit of one hundred miles a day was set. They ignored it, and when hauled into court, paid their fines and continued to defy the regulations. Perhaps the greatest offender was the Six-Day Line—six days from New York City to Buffalo. It was death on horses, but oftener than not the boats got through on schedule.

When the state legislature was called into session in January, 1824, the most controversial measure before it was another appropriation of $1,000,000 for the canal fund. The Democratic New York City delegation was solidly against it. But even with the scattered support they were able to pick up from dissidents in both parties, it was obvious that they could not muster the votes necessary to deny the money, so they shifted their attack from the Erie to Clinton personally, charging not only that under his stewardship canal funds were being wasted but also that he was misappropriating public money for his own aggrandizement.

Clinton turned to the people for vindication, and on receiving an outpouring of confidence, challenged his detractors to prove their case and remove him from office. The canal appropriation was voted, but on April 12, on a concurrent resolution of the two houses, De Witt Clinton was deposed as canal commissioner.

Only men betrayed by their corrosive envy and stupidity could have lent themselves to such an enterprise. Across the state, news of what they had done was greeted with cries of outrage. Even a considerable percentage of the anti-Clinton partisan press condemned what had occurred.

But Clinton was out—cut down when the successful conclusion of the great project to which he had devoted himself for seven years was almost in sight.

17

The Wedding of the Waters

WITH CLINTON'S REMOVAL, Myron Holley took over as acting head of the Canal Commission, as well as continuing as its treasurer, a position he had filled from the time the commission was first organized. Since he was Clinton's close friend and staunch supporter, one may wonder why he too was not dismissed. Perhaps for two reasons: he had demonstrated that he was an efficient public servant, and—more importantly—he was not politically ambitious.

Later, when the success of the completed Erie Canal was no longer a subject of debate, Clinton found occasion to say of Myron Holley: "He devoted his whole time and attention, mind and body to the canal." With such a rapport between the two men, there can be no doubt that, unofficially, Clinton continued to direct the building of the Erie. As for the political miscreants who had tried to humble and destroy him, they were soon made aware of the calamity they had brought down on themselves. All over the state in this election year (1824), there arose a popular demand, irrespective of party lines, that Clinton again be made Governor.

The fall elections were still months away when the channel from Brockport to Lockport was completed in July. Ground had been broken at Buffalo in August of the previous year. Work had been progressing on schedule and it was estimated that less than forty miles of channel remained to be dug on the entire length of the Erie, excluding the cutting at Lockport. Already improvements were being

The original Erie Canal at Frankfort, New York.

Boat line—Rochester to Albany.
Author's collection.

made on what was considered the completed canal: hydrostatic locks were being installed at Troy, Utica and Syracuse; the Niagara River lift-lock was improved; a number of new feeders were made ready; and the Salina side-cut was connected with Onondaga Lake. Meanwhile two thousand Irishmen inched forward day after day through the solid rock of the Niagara Escarpment between Buffalo and Lockport. Although their tools were still primitive, years of experience had made them expert in the use of blasting powder. With the coming of cold weather, accidents occurred more frequently. But the Irish pressed on, determined to finish the job before the season ended, and on October 26, 1824, they broke through the last barrier.

It had been a year of solid achievement for the Erie, coupled with the amazing growth of the towns along the canal. Rochester had become a thriving city of four thousand, Syracuse was growing almost as fast, Rome, Utica and a dozen other villages were emerging as prosperous towns.[1] The great central valleys of the Mohawk and Genesee that had been semi-wilderness a short time before were being cleared, with land values doubling and tripling almost overnight. So much timber was being rafted down the Erie that jams at various locks were a daily occurrence, oncoming traffic being delayed for hours.

In the November elections De Witt Clinton was swept into office with a majority of sixteen thousand votes, the largest any gubernatorial candidate had

Clinton Square, Syracuse, 1870.

ever received. Even New York City, which had so long opposed the building of the canal, responded to the increased prosperity it was deriving from the Erie by voting for him.

During the past season, according to local newspaper accounts, it had been an almost daily sight to see long lines of sixty to seventy boats impatiently awaiting their turn to get through the chain of locks between Schenectady and Albany. We have no reason to doubt it, but credulity ceases when we are told that "10,000 boats had passed the junction within the season." On examination, however, the statement is found to be more misleading than absurd. It does not say that ten thousand *different* boats had been engaged in the canal trade the past year; only that up to that number had passed a given point during the season. Since there must have been as many departures from the Albany basin as arrivals, and most of the freighters, even from as far west as Rochester, made as many as six round trips a year, the number of different boats using the canal would be about a thousand. That conclusion is justified by the fact that the tolls collected on the Erie and Champlain canals for 1824 amounted to slightly more than $300,000.

That the annual earnings would be enormously increased with completion of the Erie could not be doubted. The optimism of the moment touched off a reckless wave of enthusiasm for building a system of state-owned branch canals. The proponents of the plan were motivated by the same selfish, regional interests that two years later were to damn Pennsylvania's scheme of state-owned branch canals, as we have seen. But not having that example to guide it, the legislature passed what became known as the "Great Canal Act," authorizing the surveying and estimating of seventeen canal routes.

The story of the branch canals can be told later; 1825 was the year of the Erie. How it had grown as a corporate entity can be judged by the fact that when the season opened it had on its payroll—exclusive of construction workers— twenty-six hundred men: district superintendents, clerks, auditors, toll collectors, lock tenders and repairmen.

In June the lock gates at Black Rock were opened and Lake Erie water admitted into the western section of the canal for the first time. Two weeks later, with a host of notables in attendance, including Governor Clinton, Commissioner Holley, Chief Engineer Wright and Nathan Roberts, whose genius was about to be tested, the capstone of that remarkable feat of engineering, the famous Lockport "Fives" or "Combines"—the double chain of locks, five for ascending and five for descending traffic—were laid with solemn Masonic ceremonies.

Clinton had defaulted on his original promise that the canal would be completed in 1823. He was now under the obligation of fulfilling his second promise that it would be open to navigation from the Great Lakes to the Hudson in 1825. It must have been a peculiarly poignant moment for him as he gazed at the rocky bulk of the growing escarpment over which Roberts was to take the Erie, wondering, no doubt, if it could be done in the few months that remained, or if he would be forced to recant a second time.

As the Irish dug and blasted their way forward, several hundred carpenters and stonemasons moved in behind them. By the middle of September the work was so far advanced that Roberts advised Clinton, in Albany, that the end was in sight; work would remain to be done, but boats would be able to pass through the Combines before the end of October.

Plans for celebrating the opening of the Erie Canal with a mammoth pageant

An improved weighlock on the Erie Canal.

Opening celebration along the Hudson. Note the cannon being fired in salute. *From Whitford, History of the Canal System of the State of New York.*

such as America had never previously witnessed had been completed but held in abeyance until now. With the successful conclusion of the construction at Lockport in sight, Clinton disclosed to the public the details of the forthcoming jubilee. It created an excitement such as the state had never known. Nowhere was the response more enthusiastic than in New York City, where the triumphant voyage of a fleet of canal boats from faraway Lake Erie to the Atlantic Ocean was to end. Money was appropriated for a great gala in New York harbor, a grand ball, parade and mammoth display of fireworks. Old political feuds were laid aside temporarily. It was Clinton's hour, and no one was brash enough to dispute it.

He and his official party left the capital on October 15, bound for Buffalo. After spending the night at Schenectady, they boarded a flotilla of boats, the finest packets on the Erie. They were greeted at every stop by throngs of villagers and country people. Clinton responded with the required oratory.

Due to arrive in Buffalo, where the party was to be honored with a banquet and such niceties as the little town (now boasting a population of five hundred) could arrange, on October 25, although a day behind schedule in reaching Lockport, they were lifted over the Combines without difficulty and cast their lines ashore at Buffalo that evening.[2]

After a celebration that lasted most of the night, the visitors sat down to a hearty six-thirty breakfast which, it is reported, some were in no condition to enjoy. The delegation then boarded the boats, which overnight had been gaily decorated with flags and bunting and maneuvered into the position they were to maintain on the long return voyage down the canal. Clinton's flagship, the *Seneca Chief*, drawn by four gray horses hitched tandem and under the guidance of two sixteen-year-old hoggees, had been given the place of honor. In line behind the *Seneca Chief* were the *Superior, Commodore Perry, Buffalo* and *Lion of the West*.

On shore, a crowd of more than a thousand people was assembled, including the Black Rock band and militia, and many visitors who had come a long way to be on hand for this historic occasion. Clinton was assisted to the roof of the *Seneca Chief's* cabin, and raising his hands for silence, announced that "Of this day, October 26, 1825, the Grand Erie Canal is declared open to navigation for its entire course, from Buffalo to Albany. And may it serve the noble purpose for which the citizens of this State have built it."

The militia punctuated the cheering with a volley of musketry, the band played, and in response to the sulfurous urging of the youthful hoggees, the horses threw their weight against the breeching and the *Seneca Chief* began its long journey to salt water.

A short distance away, a cannon boomed. Its hoarse voice had scarcely died away when it was echoed from ten miles down the canal. These were the opening blasts of a grandiose plan whereby news of the official opening of the Erie could be flashed quickly down the canal and the Hudson to New York City. At intervals of eight to ten miles—depending on the intervening terrain—cannon had been

The *Wenonah,* out of Buffalo, on the Erie Canal near Little Falls, New York. *Courtesy Buffalo and Erie County Historical Society.*

Notorious Side—cut at Troy. *From Whitford,* History of
the Canal System of the State of New York.

placed in hearing distance of each other. On being alerted by the sound of distant
booming to the west, each unit fired in turn, speeding the news onward.

The cannon employed in this spectacular between Buffalo and Port Byron
were ancient pieces borrowed from the Presque Isle Navy Yard and were from
Commodore Perry's former ships and the British vessels he had captured in the
Battle of Lake Erie in 1813. At Weedsport, one of the pieces exploded as it was
fired, fatally injuring the two amateur artillerymen.

It is said to have taken the news only eighty minutes to travel from Buffalo
to the Battery. After all the forts in the harbor had joined in a thunderous salute,
news that New York City had heard the tidings was flashed back to Buffalo by
the same means it had been received.

The Opening of the Erie Canal by Charles Yardley
Turner, c. 1905. *Courtesy New York Public Library
Picture Collection.*

As the little flotilla led by the *Seneca Chief* proceeded, it was greeted with an outpouring of pride and enthusiasm such as Clinton could not have anticipated. For seven years he had exhausted himself on behalf of the Erie. Standing on the deck of the *Seneca Chief,* receiving the cheers of the crowds that greeted him at every stop, feeling it was his duty to respond with a few remarks, weary though he was, he must have felt well repaid for the slings and arrows of abuse and strife he had suffered on behalf of the canal. But the struggle had taken more out of him than he realized; he had only two years and several months to live.

At Albany there was a great demonstration; steamboats, their flags flying, were waiting to tow the boats down the Hudson to New York City, where they arrived on November 4, the welcoming ceremonies continuing until the night of November 7, with a "great illumination and fireworks display at City Hall."

Cadwallader D. Colden had been engaged by the city fathers to write an appropriate memoir of the festivities. Of the arrival of the Clinton party in New York harbor, he reported:

> The Aquatic display transcended all anticipations, twenty-nine steam-boats, gorgeously dressed, with barges, ships, pilot-boats, canal-boats, and the boats of the Whitehall firemen, conveying thousands of ladies and gentle-men, presented a scene which can not be described. Add to this, the reflections which arise from the beauty and extent of our Bay—the unusual calm-ness and mildness of the day—the splendid manner in which all the shipping in the harbour were dressed, and the movement of the whole flotilla. Regu-lated by previously arranged signals, the fleet were thrown at pleasure, into squadrons or line, into curves or circles. The whole appeared to move as by magic.[3]

The *Seneca Chief* was escorted to Sandy Hook, where the celebrated "Wedding of the Waters," painted by many artists, took place. Before leaving Buffalo, two small, finely made cedar kegs, painted patriotically and filled with Lake Erie water, had been placed in the governor's quarters aboard his flagship. At Sandy Hook they were brought on deck, and with solemn ceremonies he poured their contents into the ocean to symbolize the union of Lake Erie with the Atlantic—or to put it in Clinton's words, "to indicate and commemorate the navigable communication which has been established between our Mediterranean Sea and the Atlantic Ocean."

Most accounts of what transpired at Sandy Hook are given additional color by relating that Dr. Samuel Augustus Mitchell, already at thirty-three America's foremost geographer, contributed to the symbolism of the occasion by emptying into the Atlantic the contents of a dozen vials of water gathered from the world's great rivers.[4]

Returning to the city, the official party was treated to a sumptuous banquet, followed by gala balls at the various armories. Upwards of thirty thousand out-of-town visitors had come to New York for the celebration. They were very much in evidence as they lined lower Broadway the following day for the monster parade, with its two hundred floats, cultural societies, bands and military units. That evening they thronged the vicinity of City Hall for the fireworks display.

Modern canal boat built to navigate the Great Lakes.

New York City had outdone itself. Handsome souvenir medals had been struck and were presented to President John Quincy Adams, former Presidents John Adams, Jefferson, Madison and Monroe, and to the elderly Marquis de Lafayette, then visiting the United States. Replicas of Governor Clinton, the *Seneca Chief* and the Lockport "Fives" were stamped on dishes, pottery, neckties and silk handkerchiefs that were eagerly snapped up by the public.

Philadelphia, alarmed that with its rapid growth upstart New York City was bent on displacing it as the country's great metropolis, regarded the jubilee with a jaundiced eye. "New York has celebrated the completion of the Erie Canal Canal with excess pomp and ceremony remindful of the days of ancient Rome," observed a Philadelphia journal. "Obviously the success or failure of the Erie will greatly affect the future of Pennsylvania's proposed system of canals. We shall await the outcome with interest and, hopefully, be guided accordingly."

The Erie—the Grand Western Canal—kept no one in doubt very long. "Even its most ardent advocates," says Alvin Harlow, "did not dream of the results so quickly to flow from it. Not only was it the most successful of all American canals, it was the catalyst that spurred other States into building canals and internal improvements." Colonel William L. Stone summed it up in words that epitomized the story of the Erie: "They [the builders of the Erie] have built the longest canal, in the least time, with the least experience, for the least money, and to the greatest public benefit." [5]

When the figures for 1825 were tabulated, they revealed that the tolls collected on the Erie that year amounted to $495,000—more than the interest on its indebtedness; that 13,110 boats and rafts had passed to and fro on the junction canal between Watervliet and Albany; "that 40,000 persons had passed Utica on freight and packet boats during the season; a daily average of forty-two boats, arks and cribs."

18

Canal Mania and the Railroads

LOOKING BACK FROM THIS DISTANCE—approximately 150 years—the immediate success and prosperity of the Erie Canal would seem to have been not only predictable but unavoidable. At the time of its actual completion, which did not occur until 1827, there were, according to state figures, four thousand miles of roads in operation in New York. That included unimproved dirt roads and improved, rock-ballasted turnpikes on which tolls were collected. The overall figure sounds impressive, but less than half of it pertained to highways north and west of Albany in what was commonly referred to as "Erie Country."

In that wide, thinly settled region there were only four or five highways (using that term loosely) that were of major importance. The most heavily used was the Mohawk Turnpike, from Albany to Schenectady, where it continued as the Utica Turnpike for sixty-nine miles. Just south of Syracuse it joined the Great Western Turnpike, the route of which is followed today by U.S. 20. "This major artery," says Joseph A. Durrenberger, "carried the traveler through the beautiful Finger Lake villages from Skaneateles to Geneva and Canandaigua and continued west through Broomfield, Avon, Le Roy, and Batavia to Buffalo." [1]

South of the Mohawk route, the Cherry Valley Turnpike followed a westward path through Cooperstown and Sherburne. Another important road ran northward from Syracuse to Three Points and on down the Oswego River to Lake

Ontario. An alternative course left the Great Western at Auburn and continued north to the falls of the Genesee at Rochester, from where it cut into the famous Ridge Road to Lewiston and the lake.

The Susquehanna-Bath Turnpike must be included in any listing of the great roads of the period. It reached Bath from Albany by way of Ithaca, opening Genesee Valley trade with both eastern New York State and Pennsylvania.

The New York State salt tax, minimal when granted to the Erie Canal, now began to run into the thousands of dollars. Coupled with the tolls it was collecting (they had zoomed to better than $600,000 in 1826) the canal was becoming the state's greatest money-maker. With prosperity in evidence everywhere, it might have been expected that demands for more and better roads would confront the legislature. But there was little talk of roads; what every section of the state wanted was a canal.

Routes for seventeen lateral canals had been surveyed and rough estimates rendered of the cost of building them. Half were found to be impracticable. But seven routes were reported approved. Since the state dared not appear to favor one section over another, it resulted in a reckless trading of votes in the legislature: "You vote for my project and I'll vote for yours." The outcome was what might have been expected; funds were voted for all seven canals and construction was ordered to be begun at once—with the Erie and the Champlain canals carrying the financial load for all.

Heading the list, and by far the most controversial venture, was the Oswego Canal—running from Three Rivers Point at the head of Oneida Lake, for thirty-eight miles down the Oswego River to the town of Oswego on Lake Ontario. However, it proved to be the most profitable of all New York State's lateral canals.

In addition to the Oswego, the Black River Canal, the Chenango, Chemung, the Cayuga and Seneca, the Genesee Valley, and the short (eight miles) Crooked Lake (today's Keuka Lake) were authorized. Clinton had opposed the multiple program on the grounds that it would place a dangerous strain on the state's economy. However, up to his death in February, 1828, he strongly recommended the building of the Black River Canal, not because he expected it to be more than moderately profitable, but for its water, which, in the years to come, the Erie was surely going to need.

That farsighted prediction was realized soon after the enlargement of the Erie in 1836, when on the Long Level its very existence depended largely on the Black River feeder.

Black River had its origin in the high, rugged hills some miles to the east of Rome. By the time it found its way to Booneville, where it struck out for the north and today's towns of Port Leyden, Lyons Falls, Lowville and Carthage, it was a sizable stream. The Enabling Act for the Black River Canal called for construction of the waterway from Rome to Carthage. No provision was made for extending navigation to Lake Ontario, but that was hardly necessary, for Black River obligingly turned westward, circling around Watertown, and reached the lake at Sackets Harbor. (Sackett's Harbor, if you prefer the old spelling).

If the public is better acquainted with the Black River Canal than with some others of far greater importance, it is due almost entirely to Walter Edmonds, who introduced it to literature. Unquestionably Black River canalers were a singular lot, engaged in the pedestrian job of conveying mixed cargos of farm products to market, and picking up whatever they could for the return trip.

Rome was the capital of their world, and they seldom got farther away from it than Utica to the south or Syracuse in the other direction.

The boats were small, seldom more than twenty-five tons. The captain-owner, if married, spent most of his time on the towpath, driving the horses, while his wife doubled as cook and helmsman. When a captain-owner felt it necessary to hire on a crew, he took on one man, never more, whose duties were such as were required.

It had long been the complaint of the southern tier of counties on the Pennsylvania border that they got less than was due them from the political overlords in Albany. When the Great Canal Act became law in 1825, they united in demanding not only a connection with the Erie but also a waterway to meet the canal that neighboring Pennsylvania was to build up the north branch of the Susquehanna, and which would open the way to the anthracite fields of that state.

It was a persuasive argument and resulted in the building of the Chenango Canal, ninety-seven miles down the Chenango Valley from Binghamton to Utica, and the Chemung-Seneca Lake Canal from the head of Seneca Lake to the Che-

Picturesque but unprofitable Seneca Lake Canal in central New York State. *Courtesy New York Public Library Picture Collection.*

mung River, with Elmira its southern terminus, a distance of twenty-three miles. The latter was cheaply built, with wooden locks and was in need of constant repair from the time of its completion in 1833 to its abandonment in 1877. Although enjoying several prosperous years soon after it was opened to navigation, the arrival of the Erie Railroad ended its usefulness.

The Chenango Canal, winding through some of the most beautiful country in New York State, had a similar history of frustration. Although more than a million and a half dollars were poured into the project, parts of it were never used. Nor did the thousands of tons of Pennsylvania anthracite that were to be its bread and butter ever materialize, for the little Junction Canal, with which it was to connect at Elmira, remained unbuilt for years.

The Junction Canal, an estimated ten to twelve miles overall, was authorized along with the other laterals when the state embarked on its senseless building program, but the money for its construction was withheld year after year. When, four years after Pennsylvania had pushed its North Branch Canal up to the state line, there was still no action from Albany, the businessmen of Elmira organized a private corporation to construct the Junction Canal. Money was raised and the canal opened to navigation. For twelve years it operated at a profit, but then the railroads grabbed the business. The final blow came when the North Branch was destroyed by a monstrous flood and Pennsylvania announced that it would not rebuild.

When New York embarked on its lateral canal campaign, a great deal of the impetus came from assemblymen and senators representing the Genesee Valley counties, which was a rich and comparatively heavily populated section of the state. That it would be one of those favored regions to be given a canal was taken for granted. Although the Genesee Valley Canal was one of the first to be approved, funds for building it were not allocated until 1836. Perhaps the delay was due in some measure to the grandiose plans that had been formulated for extending it to the Allegheny River and a connection with the Pennsylvania canals. Construction got under way in 1837. A more unpropitious time could not have been chosen, for 1837 was to go down in history as the year of the great financial panic that swept the country from end to end.

In addition to the extravagant program of lateral canals in which New York was engaged, the enlargement of the Erie had been ordered in 1836, calling for a channel of 70 feet at water level, 52½ feet at bottom, with a depth throughout of 7 feet. Of course, that meant rebuilding all locks. Little thought appears to have been given to how much these improvements would cost. In the July accounting for that year the earnings of the Erie and the Champlain Canal had been sufficient to pay off their remaining indebtedness, amounting to more than $3,000,000. Seemingly political leaders were of the opinion that there was no limit to the amount of money the Erie could produce.

Certainly it was the once despised "Clinton's Ditch" that kept the state from raising taxes as the panic tightened its grip on the economy. But the depression that followed in 1839 was too heavy a load for the Erie to bear. The state was forced to borrow millions and finally, in 1842, to declare a stop-work order on all public improvements.

The Genesee Valley Canal had to wait, along with everything else. Construction was not resumed until 1847. It had been fifteen years since the first packet boat had passed up the canal as far as Mount Morris, fifty miles from Rochester.

Another nine years elapsed before a boat reached Olean. It was much too late to be of any importance; the Erie Railroad, building across western New York, had put its rails through Olean and reached its terminus at Dunkirk, on Lake Erie. Obviously no sound reason could be advanced for spending more money on the project, which had already accounted for $5,500,000. But wrongheadedness prevailed and the Genesee Valley Canal was pushed on to a natural connection with the Allegheny River in 1862. The Civil War years were its busiest. However, when abandoned in 1878, its long record was a sorry one, tolls collected amounting to only $850,000, or roughly one-seventh of the cost of construction.

More fortunate was the short canal connecting Seneca and Cayuga lakes that the state had authorized in 1826. Although only twenty miles in length, it served a useful local purpose, which was enhanced when the eight-mile Crooked Lake Canal was dug from Dresden, on the western shore of Seneca Lake, to Crooked (now Keuka) Lake. Its principal business was with Ithaca. Every year at the approach of winter dozens of small, grimy boats headed for Ithaca from the various canals to tie up for the season. The boaters who gathered there supplied Grace Miller White with the characters for her novel *Tess of the Storm Country*, which marked the first appearance of canal folk in American fiction, twenty years before Edmonds published *Rome Haul*.[2]

Although through navigation of the Erie Canal became an accomplished fact in 1825, in the following thirty years and more there were only two protracted periods when it was not being altered and improved in one place or another. Work stopped for the first time in 1832, the year of the great cholera epidemic that swept upper New York State, and again in 1842 when the "stop law" brought matters to a halt.

The cholera epidemic first appeared in Montreal and quickly spread across the border. Rochester and Syracuse were hard hit. The contagion then spread westward along the Erie, leaving little doubt that the passing boats were carrying the virus. Medical teams sent into the area by the state clamped an embargo on all navigation, and with the limited facilities available managed to stamp out the epidemic. But not before three thousand people had died.

When a knowledgeable New York State canal buff speaks of the enlargement of the Old Erie, he is referring either to the first enlargement or the second, and not by any chance to the twentieth-century Barge Canal, with its motorized boats and tankers capable of carrying a cargo of a thousand tons, which bears only a faint relationship to the Old Erie. The building of the Barge Canal at a cost of $155,000,000 is a story in itself, but a canal on which there is no towpath and boats are power-driven is too modern to be included in this narrative.

It is with the second enlargement of the original Erie that the canal historian must deal. It began in 1837 and, after dragging along for twenty-five years, was completed in 1862. As finished, it was 13½ miles shorter than the canal Clinton built; eleven of the original eighty-five locks had been eliminated, reducing the number to seventy-four. But the new locks were so much longer than the old that the net saving in lockage was slight. Shaving 13½ miles off the overall length of the canal was largely accomplished by increased river navigation of the Mohawk between Little Falls and Utica.

With the opening of the Grand Western Canal, Buffalo had become the busiest port on the Great Lakes, a predominance it was to maintain for many years. By steamer and schooner came the produce of the major portion of the

Midwest. Detroit, its only rival, was dependent on Buffalo for its market; Cleveland and Toledo were as yet only villages. Departing from Buffalo were the thousands of settlers and immigrants bound for Ohio, Indiana, Illinois and Michigan.

By the late 1820s an average of nine hundred people were arriving daily at Buffalo bound for the West. The fast packets accounted for less than a third of that number; the majority traveled less expensively aboard the "line" or small mixed-freight boats. (That this figure of nine hundred is not an exaggeration is proved by the reported arrivals at Detroit, which, of course, did not get all of the newcomers who crossed Lake Erie.)

The population of Ohio was increasing at a miraculous rate. It had long since passed the half-million mark.[3] Its pioneers had got there by one of three ways: wagon, the Erie Canal or the Ohio River from Pittsburgh. No railroad had crossed the Alleghenies as yet, nor would until 1850, some twenty years later. What was true of Ohio was true in lesser degree of the other states that had been carved out of the former Northwest Territory.

Except for that part of Ohio bordering on the Ohio River, it cannot be denied that the rapid settling of that vast territory extending northward from Columbus to the Great Lakes was traceable to the Erie Canal, nor that the canal provided access to its only, though distant, market.

When a public utility such as the Erie becomes so prosperous that it voluntarily reduces its rates or tolls, some reduction in its annual earnings may be expected. Conversely, when rates are raised, it is for the definite purpose of increasing revenue. Seemingly this sort of bookkeeping did not apply to the Erie. In 1832 it reduced its tolls (partly, it was said, to ward off the threat of railroad competition) but without any consequent decrease in earnings. On the contrary, earnings increased. And they did so again in 1837, the year of the panic, although new and higher tariffs were imposed. The explanation was simply that in good times and bad the Erie was overwhelmed by more business than it could accommodate, even with the enlargements and improvements that were constantly under way.

"Having taken your position on one of the numerous bridges, it is an impressive sight to gaze up and down the canal," wrote Colonel Stone. "In either direction, as far as the eye can see, long lines of boats can be observed. By night, their flickering head lamps give the impression of swarms of fireflies." [4]

Berm and towpath were frequently damaged by snakes and burrowing rodents, the most troublesome of which was the muskrat. They were so numerous on the Delaware and Hudson Canal that the company paid a bounty on them. When a captain saw a break developing, it was his duty to pass word along to the nearest lockmaster at once. Traffic stopped and way was made for the repair ("hurry-up") boat that was dispatched to the scene of the impending break.

Holes made by burrowing animals were not destructive in themselves, but during a deluge, water poured into them. The surrounding soil, not yet firmly settled, began to crumble and wash away. A minor break of that nature, if caught in time, could quickly be mended by packing the opening with hay and rocks, and then plastering it with clay.[5]

The proprietors of the first canal ever dug must have realized that if traffic were not controlled the result would be chaos. Accordingly, regulations were promulgated and enforced. Through the centuries they became the basic code by which all canals functioned, including the Erie. But, in addition, the Erie produced its own code of operation. It refuted the idea held by many that any man could put a boat on the canal free of restrictions and operate it as he pleased. To the contrary, he first had to register his boat, naming the owner and certifying its weight when empty.[6] He had to describe its color or colors (and some colors were restricted), and he was required to paint the name of his boat in letters at least five inches high across its stern. He had to be acquainted with the procedures to be followed on entering and leaving a lock. And he had to subscribe to the Erie policy that gave passenger packets the right of way over other craft. There was much more. He could read the Rules and Regulations at his leisure, for they were posted in every lockmaster's office.

In 1832, with upwards of twelve hundred boats on the Erie, it was felt that the canal was reaching the limits of its capacity to move them. In 1842, ten years later, when the stop-work act became law, the number had increased to more than two thousand, of which sixty-seven (less than 3 percent) were packets, 17 percent line boats and the balance classified as freighters, bullhead boats, arks and scows. These figures not only disclosed what a small fraction of the Erie's income was being derived from the privileged packets but gave weight to the argument of the line-boat companies and their allies that the packets should be ruled off the canal.

During the years when the Grand Western Canal was under construction, and for some years thereafter, it had never been the contention of the builders that the passenger business would supply an important share of the revenue necessary to make it profitable. The feeling was, however, that fast passenger packets were needed and would serve a useful purpose. Certainly the public was beguiled by the novel prospect of being able to travel in comfort from Albany to Buffalo by canal.

Their impressions of the discomforts, rather than comforts, of traveling by canal packet have been left us by such notables as Charles Dickens, Fanny Kemble, Mrs. Trollope and Harriet Beecher Stowe. They will be considered later.

It was only on the Pennsylvania Main Line and the Erie that the packets played an important role. By the middle thirties they were doing a capacity business and the line boats and other freighters had to accommodate to them. A campaign to do away with the packets was launched in New York State. When that failed, it was proposed to limit the packets to two a day, one from Albany and one from Buffalo. The press cried that down; the public wanted the packets and must have them. But a competitior—not as yet taken seriously—was closing in on the packets and in a few years was to consign them to oblivion. As you have surmised, it was the railroads.

Almost unnoticed the state of New York had chartered its first railroad back in 1826—the little Mohawk and Hudson, which was to run between Albany and Schenectady. "But," as one source says, "the public was not yet convinced that railroads would ever amount to anything. There was so little interest in the Mohawk and Hudson that the promotors could not give their stock away."

In the late 1820s very few Americans regarded the railroads as a potential threat to the supremacy of the canals. They were still in the experimental stage, with pioneer engineers and inventors such as Peter Cooper, John Stevens and Mathias Baldwin striving to perfect their versions of what a steam locomotive should be. Suddenly, however, railroads were springing up everywhere like mushrooms after a rain.

The De Witt Clinton train, Schenectady, New York.

The Mohawk and Hudson laid its strap-iron rails from Albany to Schenectady in 1834. Up the river from the Cooper Iron Works in New York City came its first locomotive, named ironically the *De Witt Clinton*, and three ornate "coaches," which were coupled together. The latter made a brave showing, as they should, for they were the last word in stagecoaches, fitted out with flange wheels to keep them on the track.

The *De Witt Clinton* train steamed along at fifteen miles an hour on the level, but the locomotive often had difficulty getting over a hill. When that occurred, passengers got out and pushed, dodging sparks from the engine as it steamed along.

By 1847 there were no fewer than ten independent railroads operating between Albany and Buffalo. Strung out along the Erie Canal, they were fiercely jealous of one another and maintained their own terminals when they could have saved money by sharing such facilities, not to mention the convenience it would have been to their passengers. Each kept the departure time of its trains a secret.

Theoretically, one could go from Albany to Buffalo by rail, but for two short stretches that had to be made by stagecoach. But, by all accounts, it must have been a wretched journey. Adding up the running time as published in the train guides of this mishmash of railroads, a "fast express" would convey a traveler from Albany to Buffalo in approximately fourteen hours. Anyone who set out with that figure in mind quickly discovered that he was the wretched victim of some horrendous miscalculations.

In *When the Railroads Were New,* Charles F. Carter relates the experience of one such traveler, and judged by other accounts it appears to have been the rule rather than the exception. Says Carter:

He got out of Albany on Saturday evening, arriving at Utica at 11:30 P.M. [some sixty miles in five hours, occasioned no doubt by a change of trains and long delay at Schenectady], where the train drowsed on a side-track until 2 A.M. Sunday, the passengers meanwhile sitting in the stuffy little coaches in a gloom made only the deeper by one candle at each end of the car. Nobody, neither conductor, brakeman nor station master, would or could tell the travelers when they might be taken further. Then, two hours after midnight, there was a bustle in the yards, a locomotive was coupled on and away [we] went, to arrive three hours later in Syracuse.

In Syracuse the passengers were in for a real wait. Throughout the morning, then the afternoon, together making up as long a day as the travelers had ever known, they attempted every little while to learn if and when they should proceed. Nobody knew, and nobody, among the railroad men, seemed to care, or give it a second's thought. So, for better than twelve hours the now worn travelers sat around in the cars, not daring to leave lest some whim of the railroadmaster set the train in motion. At half-past five in the afternoon the journey was resumed. The train arrived in Rochester (another 50 miles) in time to hear the clock strike midnight, and in Rochester it remained for six hours. This delay semed to be for lack of a suitable locomotive. At last an engine was discovered that could haul the train to Buffalo. Distance: 290 miles. Time: 38½ hours.[7]

Commodore Vanderbilt, the first of our great monopolists, is often mistakenly credited with having amalgamated the wheezy little roads paralleling the Erie Canal and bringing forth with his magic wand the great New York Central Railroad. The truth is that when the Commodore decided that he wanted the New York Central and meant to have it by fair means or foul, it was a profitable, well-organized property, with new trackage and new rolling stock, heavily capitalized (mostly Albany money) and expertly managed. The genius who had put it together and made it healthy was Erastus Corning, the wealthy Albany politician, nail manufacturer and up-and-coming monopolist in his own right.

It was Corning who got the crippling restriction limiting railroads to carrying freight only during the closed season on the canals (about five months) lifted. Not stopping there, he pushed a bill through the legislature freeing railroads from tolls on the freight they carried. These changes, coupled with through, fast, comfortable trains (for the period, that is), enabled the New York Central to take over the passenger trade on the Erie Canal and the packets disappeared.

As the years passed, the New York Central, under the able guidance of President Erastus Corning, waxed richer and richer, which did not go unnoticed by Cornelius Vanderbilt. The railroad was forwarding its passengers and freight down the Hudson River to New York City by several boat lines that it had under contract, which was an unsatisfactory arrangement, for when the river froze over, navigation necessarily stopped for several months. To accommodate their growing business, Vanderbilt foresaw that the Central would eventually have to abandon the boats and put its rails down along the Hudson River into New York. That project held some interesting possibilities, and he set out to turn them to his advantage.

He owned a controlling interest in the New York and Harlem Railroad, with its priceless right of way down through Manhattan Island to City Hall. Another road, the New York and Hudson, had been inching northward along the east bank of the river for a dozen years and had reached Poughkeepsie, from where, over leased tracks, it got to Troy. Arming himself for the confrontation with the Central, the Commodore bought control of the New York and Hudson and reaped a quick profit by allowing it to run its trains into lower Manhattan over Harlem tracks—naturally on his terms. He was now ready to take on the Central.

The bitter struggle between the so-called Albany crowd (Erastus Corning, John Pruyn, Nathaniel Thayer and others) and that ruthless corsair Cornelius Vanderbilt has been called by historians one of the most meaningful chapters in the history of American railroads, even though they do not agree on the winner. Of course, the real winner was America, for out of the struggle came the New York Central and Hudson River Railroad, one of the nation's greatest rail carriers.

If it appears to be a digression to speak of railroads in a narrative devoted to the story of our old canals, it really is not, for it was the locomotive, belching sparks and smoke as it chugged across the landscape at fifteen miles an hour, that dealt canals the deathblow. As railroads improved, the canals, steamboats and turnpike companies buried their old differences and united to fight the brash upstart that was to destroy them. A campaign of newspaper ads and billboard broadsides warned the public against the death and destruction the railroads would spread throughout the land. From hundreds of pulpits eminent divines exhorted their flocks to demand that the railroads not be permitted to desecrate the Sabbath by operating their trains on Sunday. All to no avail.

Some companies took heed, but the public demanded "fast" transportation and was not deterred by the frequent accidents. In the railroad-building mania that followed, reminiscent of the canal mania of the 1830s, new roads and branch lines of the established carriers formed a network of rails, bringing competition that the canals could not meet. One after another they opened their locks for the last time and passed into oblivion. In New York State, the Chemung, Chenango, Genesee Valley, Black River and Cayuga and Seneca were abandoned. Of all the once numerous branch canals, only the Oswego and Champlain continued to function. Even the long prosperous and expertly managed Delaware and Hudson underwent a metamorphosis and became the Delaware and Hudson Railroad.

Although railroad competition made it impossible for many smaller canals to continue, it had little effect on the flourishing Erie, partly because the Grand

Western Canal was doing largely a long-haul business. Transportation charges on through freight from Albany to Buffalo, for instance, were considerably less by water than by rail. The boats were slower, but most shippers preferred the savings in money to time.

Of course, New Yorkers had come to regard the old Erie as a venerable and profitable institution. They manifested their confidence in the waterway on many occasions but never more so than when plans were broached for widening and deepening parts of the old channel and connecting them with a new waterway large enough to accommodate freighters and tankers capable of navigating the Great Lakes, even the ocean, the undertaking to cost an estimated $150,000,-000. Howls of disapproval arose when it was disclosed that the project was to be known officially as the New York State Barge Canal. Old-timers refused to accept the name and continued to refer to the great waterway as "the Erie." Today's canal buffs still do.

Stagecoach and canal boat companies unite in a campaign to warn the public against the steam cars. *From Thompson,* A Short History of American Railways.

19

The Ohio Canals

THE OHIO CANALS were the legitimate offspring of the Erie. That ribbon of water was not only largely responsible for the rapid settling of Ohio but also dominated its thinking when it became obvious that its prosperity could be assured only by establishing a market for the agricultural abundance of its farmlands and the products of its budding industry. When it was first proposed that the problem could be solved by digging a canal to connect Lake Erie with the Ohio River, the idea was widely supported. The wrangling began when it came to deciding where the canal should be located.

The eastern half of the state was the most populous and insisted that the canal leave Lake Erie at Cleveland by way of the little Cuyahoga River and proceed in a southerly direction to Columbus, the capital, and following the Scioto River, reach the Ohio at Portsmouth. Cincinnati and the western section demanded just as vehemently that the state build a canal up the Miami River, connecting Cincinnati with Hamilton, Dayton and via the Maumee with Toledo and Lake Erie.

The young state, with its lean treasury, was hardly in a position to build one canal, let alone two. The result was that nothing was done for three years. Finally $6,000 was voted by the legislature for a survey of five proposed routes for a north-south canal. James Geddes was brought over from the Erie to do the surveying. Heavily favored were what was to become the Ohio and Erie

Type of boat used on the Ohio canals.

A packet on the Miami and Erie Canal. One of the many that left the Erie Canal to escape railroad competition. *Courtesy National Archives.*

Canal and the Miami and Erie. But with public interest divided on strictly sectional lines, it was apparent that one could not be built without the other.

Since the two sides to the dispute refused to compromise their demands, the legislature took no action. It was not until former Superior Court Judge Ethan Allen Brown of Cincinnati was elected governor in 1818 that the canal question became a political issue. For years Brown had been a forceful advocate of the construction of a canal connecting the Ohio River with Lake Erie. That was the issue on which he had campaigned and been elected. In his inaugural address he devoted himself almost exclusively to the subject, pointing out for the benefit of those in the audience who felt that such a costly internal improvement could not be built without raising taxes to a prohibitive rate that the federal government owned millions of acres of public lands in Ohio, and in keeping with its policy of assisting the new states, would donate a portion of them to Ohio, which could be sold or on which funds could be borrowed.

Notwithstanding his vigor and enthusiasm, it was not until 1822 that Governor Brown succeeded in getting a Canal Commission appointed. The reports were so favorable that the legislature had to respond affirmatively. The Enabling Act, authorizing the construction of the Ohio and Erie Canal (referred to as the main canal), become law in February, 1825. It also provided for construction of the Miami-Maumee Canal from Cincinnati to Dayton, with the proviso that it would be extended to the Maumee River and Lake Erie "at some future time." Without this sop to Cincinnati and the Miami Valley farmers the legislation would have been defeated. But years were to pass before work on the promised extension got under way.

Before construction of the Ohio and Erie began, another and final survey was made by David Bates. Bates was still another graduate of Judge Benjamin Wright's "Erie school of engineering." With only several minor diversions, the route he submitted followed the course previously staked out by Geddes. It ran up the Cuyahoga River through Akron, Canton and Dresden, and after reaching the Muskingum River, headed west by way of Newark to Columbus and down the Scioto through Chillicothe to the Ohio at Portsmouth. Canton didn't want the canal, so Bates swung it eight miles to the west through Massillon.[1]

The canal commissioners were so impressed with Bates that in the late spring of 1825, with construction slated to begin on July 4, he was engaged as chief engineer. Time was to prove that their trust in him was not misplaced; when he needed advice, he sought it in the best of all places, among his old cronies of Erie days, bringing in men like Nathan S. Roberts and Canvass White to serve as consulting engineers.

It is impossible to appreciate how much the state expected from the Ohio and Erie Canal unless one remembers that in the midst of plenty a great proportion of its farm population was living on the edge of poverty, even though it was producing hard grains, pork products and hard vegetables in abundance. With wheat selling at $.25 a bushel, corn for even less and domestic flour bringing an average of only $1.95 a barrel, their plight was desperate. The canal, 308 miles long, was to change all that, opening a route to the New York market, via the lake and the Erie Canal, in one direction, and southward by way of the Ohio River to Pittsburgh and the Pennsylvania market.

The prospect was made all the rosier by the knowledge that a barrel of flour

Basin of the Miami and Erie Canal at Cincinnati.

Corn being hauled to Chillicothe on the Ohio and Erie Canal. *Courtesy National Archives.*

A wheat and corn carrier on the Ohio and Erie Canal.

could be transported via the Erie Canal to New York City for $1.80, where it would fetch $8.00. No wonder that the Ohio and Erie Canal was regarded by many as a dream come true.

Among the thousands who gathered at Licking Summit level, three miles west of Newark, for the ground-breaking on July 4 were hundreds who had driven or ridden saddleback scores of miles to be present at the ceremonies.

Like Brown, whom he had succeeded, Governor Jeremiah Morrow was a staunch supporter of the canal movement. In his determination to make the turning of the first sod a notable occasion, he had invited De Witt Clinton to be the principal speaker. Busy though he was with the completion of the Erie in sight, Clinton accepted.

Few men in America were better known, and his presence alone guaranteed the success of the occasion. He reached Cleveland with his party on the last day of June, where he was met by a welcoming committee of Ohio notables. His progress from Cleveland to Newark has been described as one continuous ovation. There, on July 4, the accepted day for such occasions, Governor Morrow, surrounded by public officials, clergymen and a regiment of militia, was waiting to greet him.

Bands played and cannon boomed as Clinton ascended the speakers' platform. He made what has been described as an eloquent address, after which, at a spot indicated by Chief Engineer Bates, he drove a spade into the earth and turned the first piece of sod. Governor Morrow then lifted the second shovelful. Others, who for political reasons could not be denied, stepped forward and took their turn. The official party then proceeded to Columbus where Clinton was tendered a banquet. He was escorted to Middletown the following day, and ground was broken for the Miami and Erie Canal.

With a minimum loss of time, construction of the mandated 60 miles of the Miami and Erie and the 308 miles of the Ohio and Erie began. Scores of contracts were let to the professional labor bosses whose Irish crews had dug New York's waterways. In addition, contracts calling for the digging of as little as half a mile of channel were given to local citizens. This was an innovation by the

Boats about to pass on the
Miami and Erie Canal.

Famous packet *St. Louis* on the Miami and Erie Canal.

The Miami and Erie Canal crossing the Great Miami River.

state for the laudable purpose of keeping a share of the money the canals were costing in the various communities in which it was being earned. To the small-town businessmen and industrious farmers who took advantage of the opportunity it was a boon, providing jobs and wages.

No account of the contribution made to the building of the Ohio and Erie by these amateur contractors fails to mention Abram Garfield of Tuscarawas County. No doubt he was typical of the hard-working, industrious pioneer farmers of his day, but his greatest claim to fame is that he was the father of James A. Garfield, born in 1831, who became the twentieth president of the United States. The senior Garfield, having taken a contract to build half a mile of channel, recruited a work force of twenty men among his neighbors, built shanties in which to house them and, turning his log cabin home into a boarding house, assigned to Mrs. Garfield the not inconsiderable chore of putting three meals a day on the table for all the workers.

By late fall upwards of two thousand men were at work north of Licking Summit. Their main camp was at Cuyahoga Falls, the site of today's Akron. As many as a fourth of them were young natives, the rest experienced Irish and German laborers brought in from New York State. Across Ohio, north of Cincinnati, work on the Miami Canal was proceeding at a similar pace.

Construction of the Ohio canals, as far as wages, time lost to bad weather and outbreaks of malaria and cholera (1833) were concerned, echoed the pattern that had obtained on the Old Erie. But the similarity ended there, for no major engineering problems, such as taming the lower Mohawk, carrying the Erie over the Genesee River and building the Lockport Fives, confronted David Bates and his eager but untried assistants. He was harassed, however, by a continuing problem of a quite different nature, and about which he could do nothing.

It had to do with the precarious financial position in which Ohio found itself as the result of its canal program, despite the generous land grants it had received from the federal government. In some quarters there was a rising demand that the canals be abandoned and the drain on the state's treasury stopped.

"Out in the field there were times when we didn't know what the next day would bring," Bates recalled later.

"We had about 3,000 men and as many as 1,500 teams of horses on the job just before the cholera swept through the camps and thinned the ranks. The contractors could not do much for their crews, being strapped for cash on account of the State holding back money that was overdue them. Two or three simply throw up their hands and quit the country, leaving their men unpaid.

If work on the Ohio Canal never ground to a halt, the credit was due almost exclusively to two of its Board of Canal Commissioners, Micajah T. Williams of Cincinnati and Alfred Kelley of Cleveland. Kelley was a lawyer with private means, which was fortunate, for he often had to dig into his pocket to cover expenses as he and Williams traveled about Ohio. They served the state almost without pay, each receiving a miserly salary of $3 per day.

A portrait of Alfred Kelley, Ohio's great canal commissioner. His indomitable spirit and financial wizardry saved the state's canal program when it tottered on the brink of collapse and bankruptcy.

On the theory that the stimulus of a gill of whisky doled out to a crew four times a day enabled a contractor to get more work out of it, labor bosses had gone along with that custom to the extent that it had long since become routine. Williams and Kelley didn't object to the whisky but to the time lost by the breaks. Knowing they would be stepping on tender toes and might precipitate a riot, they proved their mettle by ordering the practice stopped. The men growled their displeasure and threatened to strike, but after they had vented their displeasure, they went back to work, which proceeded with noticeably greater efficiency.

The Canal Commission faced one crisis after another, but Kelley was always able to find money enough somewhere to tide things over for a few weeks. He organized land sales with a circus atmosphere and conducted a lottery. In December, 1842, he went East in a desperate attempt to borrow money on state bonds. His errand appeared hopeless, for it came at a time when Indiana, Illinois and other states were defaulting on their notes. Ohio's credit was so suspect that Kelley could not borrow on it. If, after several weeks of being rebuffed, he was able to raise sufficient money to see Ohio through the current crisis, it was only by pledging his own personal resources as security.

In their historical perspective De Witt Clinton and the Erie Canal are one; that the Erie would have—or could have—been built without Clinton's dedication and perseverance must be doubted. Likewise, one man, Alfred Kelley, was largely responsible for the successful completion of the Ohio canals. Although he never received the acclaim due him and is unknown today, his financial wizardry and unwavering faith in the prosperity the canals would give the state made their completion possible. Fifteen years later the canals could not have been constructed, for Ohio was building more railroads than any other state in the Union. But before he stepped down, Kelley had the satisfaction of seeing the once worthless canal bonds quoted at par.

Cuyahoga Falls has been mentioned. There, in June of 1827, the canal boat *State of Ohio* was launched. It was small, less than fifty tons. On July 4, to the cheers of the attending multitudes, it began the first journey ever made on an Ohio canal. This section had recently been completed to Cleveland, a distance of thirty-seven miles. The drop from Akron to the lake was measured at 395 feet, requiring forty-one locks, which made it the costliest and most tedious segment on the entire length of the Ohio and Erie.

The finished section, short as it was, was declared open to navigation and in the remaining months of that season more than $1,500 in tolls were collected, which must have encouraged Alfred Kelley. The following year boats were operating between Cleveland and Coshocton, with more cargo being offered than they could accommodate. That same season the Miami was opened between Middletown and Hartwell's Basin, near Cincinnati. So much flour and other farm produce had been wagoned into Middletown in anticipation of the opening of navigation that the few boats available were able to make a dent in it only by running day and night. The weather being favorable, navigation continued until the middle of December. Cincinnati newspapers reported that in addition to other items, "58,000 barrels of flour and 1,800,000 pounds of bacon have reached this city over the Miami and Erie Canal."

Over $8,000 in tolls had been collected. It was telling evidence of the richness of the Miami Valley and the volume of business Cincinnati might expect from the canal. Pressure was put on the Canal Commission to extend it to Dayton, its present authorized terminus, the following year. Not only was that done but the Miami Canal was brought directly into Cincinnati.

To take advantage of what appeared to be a bonanza, shipyards along the Ohio began turning out a rash of canal boats, neat craft of fifty- to sixty-tons burden. If they rested comfortably on the water and had graceful lines, it undoubtedly was because they were put together by English shipwrights who knew their trade, rather than by country carpenters, samples of whose handiwork was often seen on other canals.

The first boat through from Cincinnati reached Dayton in January, 1829. Although the lower Miami Canal was open to navigation, work remained to be done and it was not completed until the end of that year. Earnings for 1829 amounted to $16,000. Shippers received higher prices for their mill and farm produce. Those sections of the upper Miami too distant from the canal to take advantage of it united in demanding that the promised northern extension to the Maumee River and Lake Erie be undertaken at once. The state was having such a desperate time finding money with which to complete the Ohio and Erie Canal that it refused to consider extending the Miami. The extension to Lake Erie was not even authorized until 1831, and two years elapsed before construction began.

In the meantime the Ohio and Erie had been completed from Cleveland to Chillicothe, as had the short branch from Lockbourne to Columbus, 260 miles. Considerable work had been done on the remaining fifty-eight miles to Portsmouth. Bates hoped that it could be hooked up with advancing construction from the north and the canal completed before the end of the year. He was proceeding according to schedule when the cholera epidemic broke out in August and brought all work virtually to a standstill. With the coming of cool weather the worst of the pestilence passed. Bates limped forward with his depleted crew. On December 1, 1832, the lock gates at Portsmouth were opened for the first time and a flotilla of waiting boats passed through into the Ohio River.

The Ohio and Erie Canal had cost $7,904,970, somewhat less than twice the estimated figure. Including the eleven-mile Columbus branch, it was eighteen miles longer than originally planned. The feeders Bates built on the Portage summit, and which proved vital, were another added expense. The canal's value to the state cannot be computed by measuring its earnings against its cost. In 1837 earnings passed the $100,000 mark and for the following ten years they never fell below that figure. In 1851, the canal's most prosperous year, they reached $432,070.

Being on the canal was the making of such towns as Akron, Massillon, Newark, Coshocton and Chillicothe. With the opening of the Ohio and Erie, commerce began flowing north and south to distant markets, increasing both the demand for and value of whatever was being produced.

In the spring of 1833, with the Ohio and Erie Canal doing a flourishing business, work was begun on the extension of the Miami to Toledo and Lake Erie. Where the state, whose accumulated indebtedness already amounted to $4,000,000, expected to find the money for building the Miami extension, estimated to cost at least $7,000,000, (the actual cost was $8,062,670) is a mystery. Possibly it expected the federal government to come to the rescue if the state failed to extricate itself from the muddle it was in. But in view of the liberal treament already received from that source it was unlikely that further aid would be forthcoming.

In 1828 Congress had made an original grant to Ohio of 500,000 acres of land to further its canal program. Later, an additional grant of 464,000 acres had been made for the exclusive benefit of the Miami and Erie Canal. From the sale of these lands the state had realized in excess of $2,000,000. Again in need of money, it engaged in some political horse trading with neighboring Indiana, which was building its famous Wabash and Erie Canal, eventually to become the longest canal dug in America, and secured another grant in return for agreeing to build the eighteen-mile link of the Wabash between the Indiana border and Defiance, Ohio, and granting to the Wabash the right to use the Miami from Defiance to the lake. All told, the federal grants totalled 1,230,000 acres—a giveaway no other state secured.

The population of Ohio nearly doubled in 1834–36. Prices were high and money was cheap. The legislators in Columbus were led astray by the wave of prosperity sweeping the land. They not only authorized construction of several state-owned canals but put money into some that were privately owned. Not stopping there, they launched a program of turnpike construction and even advanced funds to several burgeoning railroads.

Most publicized of the private canals was the Ohio and Pennsylvania. In 1836 the Ohio legislature subscribed for one-third of the original stock issue of this canal, whose purpose was to connect the Ohio and Erie Canal with the Pennsylvania Main Line Canal. With this assistance the company was able to open its channel from the Pennsylvania border to Akron in 1840. The Pennsylvania end of the canal was never built. After a losing struggle the entire project was abandoned.

Another private corporation had built a short canal from the Ohio and Erie into the Hocking Valley and its rich coal fields. The state took it over and extended the waterway down to Athens, fifty-six miles in all. It would have been a worthwhile venture but for the freshets that were in the habit of tearing down the Hocking River and putting the canal out of commission for weeks at a time. When the state finally wearied of trying to keep it in repair, it withdrew, and the Hocking Valley Railroad, another private corporation took over.[2]

Another attempt to effect a short-cut connection with the Pennsylvania canals was the seventy-three mile Sandy and Beaver Canal. Taking off from the Ohio and Erie at Bolivar, some miles below Akron, it ran directly east to the Ohio River forty miles downstream from Pittsburgh. This was a private promotion in which the state miraculously escaped becoming involved. Construction began in 1835 and was not completed until 1846, and proved to be a complete fiasco, costing its stockholders some $2,000,000.

Equally ill advised was construction of the Walhonding Branch of the Ohio and Erie, which originated at Coshocton and ran twenty-five miles northwest to nowhere. Financed by the state, it lost money consistently until abandoned.

Among other white elephants that were unloaded on the state was the short Warren County Canal, chartered in 1830 to connect the town of Lebanon with the Miami Canal at Middletown. The work was no more than half finished in 1836 when the government stepped in and expended several hundred thousand dollars to complete it.

Ohio was to pay dearly for the foolish financial antics in which it overindulged in 1835 and 1836. But while the boom lasted every section of the state wanted to get its piece of the cake. Cincinnati was no exception. It was the most populous city in Ohio and politically powerful.[3] Having been assured that Ohio and Indiana would cooperate in financing the venture, a group of Cincinnati businessmen in-

At Muskingum.

Where the Muskingum Canal met the Ohio River. *Courtesy New York Public Library Picture Collection.*

corporated the Whitewater Canal Company for the purpose of building a twenty-eight-mile-long canal to connect the city of Cincinnati with Indiana's Whitewater River and the most densely populated area of that state.

Cincinnati, which was to be the sole beneficiary, subscribed $40,000, Ohio $150,000 and Indiana $150,000. Indiana was in such financial difficulties that it could not fulfill its obligation.

The canal followed the Ohio downstream for several miles, then crossed the Great Miami River by aqueduct at Cleves, from where it pointed northwest for the Whitewater River of Indiana. Since the company was almost continuously out of funds, the work dragged on and was not completed until 1843. Completion was not followed by the rush of business that had been predicted. In fact, the Whitewater Canal was never profitable. It was severely damaged by freshets in 1846. The following spring it suffered further destruction. The canal was abandoned in 1847 and the old channel, drained and converted, became the right of way of the Cincinnati and Indianapolis Railroad.

All but one of Ohio's minor canals must be regarded as failures. The exception was the Muskingum, which cut away from the Ohio and Erie at Dresden Junction, fifteen miles below Coshocton, and followed the Muskingum River down to its mouth on the Ohio River near Marietta. Of the several projected short-

Akron, Ohio, on the Pennsylvania and Ohio Canal.

cuts to Pittsburgh and the Pennsylvania canals, it alone was profitable. Begun in 1836, it was about equally canal and river navigation. At its mouth a basin was dug for the convenience of steamboats. As a consequence the tonnage transshipped there often exceeded that registered at Portsmouth.[4]

Construction of the Miami extension had been proceeding according to schedule throughout the year, and as 1836 drew to a close, the channel had been dug as far as Troy, twenty-one miles north of Dayton. Although several important New York and Philadelphia banks failed in December, there was no widespread apprehension that bad times were ahead. But as spring advanced, signs multiplied that the nation was facing its first great financial crisis.

The panic of 1837 was at hand, and no state was more unprepared to meet it than Ohio. It was not only wallowing in debt but obligated by its agreement with the federal government to proceed with construction of the Miami and Erie as well. A temporary respite could be gained by holding back the money due the contractors who were building the canal, but only for two or three months, for they had to meet their payrolls or see their crews walk off the job.

Somehow enough funds were found to make it possible for construction to continue. Another ten miles were completed and on July 4, 1837, the first boat, all the way through from Cincinnati, reached Piqua, the seat of Miami County, and was welcomed by a cheering crowd of several thousand who had gathered for the annual Independence Day celebration.

At that time an estimated three thousand men were engaged in building the Miami Canal. In addition to the gangs of Wild Irish who had opened the channel to Piqua, other hundreds were toiling on the Maumee. Between Defiance, Ohio, and the Indiana line, the connection with the Wabash was being dug. At St. Marys, on the summit level of the Miami, construction of the Grand Reservoir, which was to be the Miami's chief source of water, was under way.

In all the camps there was growing unrest as wages remained unpaid. Only by making token payments with money borrowed from private sources were the

Wrecked aqueduct on the Miami and Erie Canal over Auglaize River.

contractors able to keep the men from throwing down their shovels and refusing to work.

By the end of the year the numbing effects of the panic receded and business began to recover from its stagnation. The state had managed to keep its notes out of the hands of speculators. Aided by a recurring interest in purchasing public lands and the increasing tolls the canals were providing, its financial position improved.

Of all the problems connected with building the Miami and Erie Canal none was more troublesome than construction of the Grand Reservoir, located in Auglaize County and extending westward into Mercer County. It was formed by two dams, twenty-five feet high, one two miles and the other nearly four miles long. Its water surface covered twenty-seven square miles, being approximately nine miles long and from two to four miles wide. It was reputed to be the largest artificial lake in the world, and was regarded with suspicion by those who feared the consequences if a break occurred, as well as hated by those who had been evicted and had to wait three to four years before they were paid for their land.[5]

The country had not fully recovered from the panic of 1837 when it was confronted by the depression of 1839. Public works were the first to feel its numbing effect; construction slowed or stopped altogether. Completion of the Miami and Erie was delayed for a year and a half, and it was not until May, 1843, that the section along the Maumee was opened to navigation and the connection with the Wabash finished. Boats began reaching Toledo from Fort Wayne and other Indiana towns, laden with grain and other farm produce. Returning, they carried salt, sugar, coffee, hardware and agricultural implements.

Ostensibly the Miami and Erie Canal was completed in 1844, but the section between the Maumee and the summit level could not be used because the Grand Reservoir was not yet finished. It was declared serviceable in the late spring of 1845. Water was let into the channel and the first boat passed through from Toledo to Dayton in June. Not including its 25 miles of navigable feeders, it was

Just a memory—an old lock
on the abandoned Miami
and Erie Canal.

248 miles long. In its first full year of operation, it returned to the state $233,527.

If you trace out the course of the Miami Canal on a modern map you will find it passing through one thriving city after another. Without exception they owed their early prosperity to the canal. It had a tenacity that few other canals exhibited. Even when the Cincinnati, Hamilton and Dayton Railroad built tracks to parallel it almost all the way, the Miami refused to surrender. As late as 1903, long after most canals had become a memory, the tolls collected on the Miami amounted to more than $70,000, even though the railroad lobbyists had persuaded the legislature to shut it out of downtown Cincinnati and reduce its wharfage rights at Toledo.

During its halcyon years fast packet service on the Miami equaled the best to be found on any eastern canal. The boats themselves were as luxurious as those seen on the Erie. In fact, some of them had plied Erie water before the railroads moved in and captured the passenger business. In their old age on the Miami they were to enjoy some prosperous years before the Iron Horse overtook them for the second time.

20

The Indiana Canals—
Half a Century of Folly

WHEN INDIANA WAS admitted to the Union in 1816, its population totaled fewer than 135,000. Four years later, the official census of 1820 could account for only 147,148—not including Indians, of whom a thousand remained. Most of the pioneers had crossed the Alleghenies and reached their new home by rafting down the Ohio River. The majority had settled in the southern section of the state; north of Terre Haute, Indiana was still frontier or what was known locally as "fringe country," with Lafayette and Fort Wayne the only settlements of any consequence.

So completely has the evidence been removed that in traveling through Indiana today it is difficult to believe that it once was a heavily forested land, with great stands of oak, maple, towering elms and sycamores. And yet there are local folklore tales of newcomers taking their first look at the so-called farmland they had bought at a distant land office and being appalled by the number and size of the trees (some six feet in circumference) that would have to be cleared before a crop could be planted.

Indiana had thousands of acres of land for sale, a gift from the federal government. To increase sales the price per acre was lowered from $2.00 to $1.78, with only one-fourth in cash being required. Sales were held at various places. They were disappointing, bringing in less than $75,000. The year's total tax levy amounted to only $33,000. The money received from these two sources was barely sufficient to cover the expenses of the state government. The voters were well acquainted with the fact that only by practicing the most rigorous economy was Indiana able to carry on. It should have been enough to silence those who were demanding that the state plunge into a program of costly internal improvements without any money in sight with which to pay for them. Instead, the clamor grew for transportation improvements that would pull Indiana out of stagnation. If other states could build canals on borrowed money, why couldn't Indiana?

In the legislature the representative from Marion County startled his associates by declaring that the best way to solve the transportation problem would be to build a north-south railroad connecting Lake Michigan with the Ohio River, which would serve all parts of the state. This was at a time when not a foot of track had been laid west of the Hudson. Nevertheless, it was seriously debated until assemblymen from the Whitewater River counties announced that they would have no part of any railroad scheme; what they wanted was a canal paralleling the Whitewater River that would give them access to the Ohio.[1]

There had been talk of this before and in 1827 a private corporation had been chartered to build what was called the Whitewater River and Fort Wayne Canal, which would have been a disaster. Fortunately nothing had come of it.

That same year the Congress had given Indiana a belt of land 2½ miles wide on either side of the channel it might build to connect the Wabash River with the Maumee. Since the navigable waters of the two rivers could not be reached without crossing a state line, the grant was meaningless unless some arrangement could be made with Ohio. The legislature mulled the matter over for two years without acting. But in 1829 representatives of the two states met in Cincinnati and reached an agreement.

In accepting the land grant from the federal government, Indiana agreed to connect the navigable waters of the Maumee with the navigable waters of the Wabash. It had been so concerned with constructing a canal through the marshes east of Fort Wayne and bypassing the falls of the Maumee that little thought had been given to what it faced in the other direction. Belatedly it realized the Wabash was not navigable short of the mouth of the Tippecanoe River, a few miles east of Lafayette. That added up to a little better than a hundred miles of canal construction.

The officials and politicians who had maneuvered so that Indiana was not committed to building a Wabash and Erie Canal were now singularly reluctant to pass the necessary legislation so that work could be begun. But after two years of teetering back and forth, doing nothing, they took the plunge, spurred on by news of the fantastic prosperity of the Erie Canal and the tolls being collected on the half-finished Ohio canals. A loan of $200,000 was authorized, which was to be placed in the hands of a Board of (three) Fund Commissioners, who were to have power to act as their judgment dictated.

The board went into action at once and succeeded in engaging Jesse Williams, an Erie Canal engineer who had been working with David Bates on the Ohio canals. Williams met the board in Fort Wayne. On Washington's birthday, 1832, ground was broken for the Wabash and Erie in the presence of the usual cheering crowd.

The earth was frozen and covered with a foot of snow. There were other aspects than the weather in which the occasion differed from the customary groundbreaking ceremonies. Never before had a canal been dedicated prior to being surveyed and estimates rendered of its probable cost. What was even more remarkable, the board, as yet, had only the haziest idea of where their waterway was to cut into the Wabash.

The Wabash River rises in Ohio and for the first sixty miles of its twisting, uncertain northwest course it is a piddling stream of little consequence, until it reaches Huntington. Although a score of Hoosier writers have glorified the Wabash in song and story it is a placid, muddy river, except when a series of freshets turns it temporarily into a rampaging, synthetic tiger.

When the French were in command of the country, they followed the Wabash across Indiana in their canoes and pirogues and turned south with it where Lafayette now stands, which carried them to Terre Haute, Vincennes and their settlements on the Ohio.

Jesse Williams wisely decided that this was the course the canal must follow, once it had got away from Fort Wayne by a viaduct over the St. Mary's River. Accompanied by several assistants, he spent the spring running an exploratory survey. When he submitted his report it was accompanied by his estimate of the cost of the improvement. It amounted to slightly more than $1,000,000.

By the end of the year a thousand men were working on the Wabash and Erie west of Fort Wayne. More money was needed, and the state succeeded in borrowing it. In the meantime, Ohio having done nothing about fulfilling its commitment on the Wabash and Maumee connection, Williams began running lines on his own authority.

As early as 1826 Fort Wayne businessmen had hired James Shriver, a local surveyor, to run lines across the marshes of the portage east of town. It was mosquito-infested country in which a man might be exposed to a dozen different varieties of fatal fevers. Shriver brushed aside the warnings and established a camp in the marshes, only to be fatally stricken within a month. Asa Moore succeeded him and met the same fate. Williams had heard these tales but they did not deter him. He spent ten days in the marshes, bothered more by rattlesnakes than anything else. He went on beyond the falls of the Maumee for several miles and saw enough of the rapids below to be convinced that the river could not be used as a channel for the canal, as the board had hoped.

By the end of the year nineteen miles of the canal, from Fort Wayne to Huntington, and two feeders were completed. In the following July boats began plying between the two towns. Less than $5,000 was collected in tolls, which was not sufficient to pay for the repairs made on the wooden aqueduct over the St. Marys River at Fort Wayne and the replacing of several locks whose gates were already warping.

When the Canal Board announced that instead of ending construction of the Wabash and Erie at the mouth of the Tippecanoe the canal would be extended an additional ten miles to Lafayette, the wisdom of doing so was not questioned—Lafayette was the most prosperous and fastest-growing town in the state. The decision to make it the western terminus of the Wabash and Erie was interpreted by the optimistic *Tippecanoe Journal* as unmistakable evidence of Lafayette's future as a great lake port, even though it was a full eighty miles from the nearest point on Lake Michigan. A canal deep enough to accommodate steamboats would solve that problem. Their assurance that the commerce of Lake Michigan would become a thriving billion-dollar business cannot be explained. At the time there were not more than half a dozen small steamers plying its waters, and the only

Scene on the Chesapeake and Ohio Canal, c. 1875. *Courtesy New York Public Library Picture Collection.*

port town that gave promise of amounting to anything was the still shabby village on the mud flats at the mouth of the Chicago River.

Needless to say, the suggested Lake Michigan–Lafayette Canal was not built nor was Lafayette's dream of becoming a lake port ever realized.

The Salamonie, a sizable river in its own right, discharged its waters into the Wabash at the village of Lagro, known familiarly to Hoosiers as the Forks of the Wabash. For months the main camp of the canal builders had been located there. By July 1, 1836, the channel was completed to that point and a short side-cut dug, in which boats could be turned around. Water was let into the summit level and everything made ready in good time for the ceremonies that were to mark the official opening of the first thirty-two miles of the Wabash and Erie Canal.

At high noon on July 4, in keeping with the canal custom, three boats from Fort Wayne, decorated with flags and bunting, reached Lagro, bearing Governor Noah Noble, his staff, the Canal Board and Fort Wayne's leading businessmen. There was the usual optimistic oratory, the speakers agreeing with Governor Noble's observation that "Indiana was on the move at last." No one mentioned the state's growing indebtedness.

The population was increasing at a rapid rate. Thousands of "movers" and German, Swiss and Scottish immigrants were being lured westward by the prospect of acquiring cheap land in Indiana. In the north, after reaching Toledo by steamer, the newcomers made their way in wagons up the Maumee Valley to Fort Wayne; in the south, steamboats set them down at one or another of the state's Ohio River ports. It augured well for the future, since a man automatically became a taxpayer as soon as he acquired land.

Undoubtedly this had something to do with the thinking of the legislators when they convened in the new capital of Indianapolis in January, 1836.[2] In line with the feeling of prosperity that was sweeping the United States, they went hog wild and put together the Mammoth Internal Improvement Bill. It was designed to give every section of the state a windfall—if not a canal or railroad, at least a turnpike. Governor Noble signed the bill on January 26, 1836, and by that rash act saddled Indiana with a debt of $13,000,000, which was to cripple and embarrass her for half a century.

But no one thought so at the time. As news of the signing of the bill spread, it touched off celebrations in town after town. Bands turned out and bonfires were lighted. It was as though the millennium had come, that by the scratch of a pen Indiana would be lifted out of its backwoods doldrums into a world of progress and opportunity. Of course, there would be an awakening, and it would be painful. But while the dream lasted, life quickened everywhere. A wave of speculation swept the state. Real estate values increased everywhere. In the large towns speculators amassed small fortunes.

The Mammoth Internal Improvement Bill provided for the appointment of a Fund Commission of six members, each from a different section as a guarantee that no one section would be favored over another. Along with being held responsible for completion of the improvements outlined in the bill, they were authorized to borrow money in the name of the state. They began by borrowing $10,000,000 at 6 percent for a period of twenty-five years.

The Internal Improvement Bill authorized the following works:

1. The Whitewater Canal; also a canal to connect the Whitewater with

the Central Canal. (If this was found to be impracticable, a connecting railroad was to be built).

    2. The Central Canal, connecting with the Wabash and Erie at Logansport, and running via Indianapolis and White River to Evansville.

    3. Extension of the Wabash and Erie from Lafayette to Terre Haute and via Eel River to the Central Canal and the Ohio River at Evansville.

    4. A railroad from Madison to Lafayette.

    5. A turnpike from New Albany to Vincennes.

    6. A railroad or turnpike from Jeffersonville to Crawfordsville.

    7. A survey made for a canal (or railroad) from Fort Wayne to Lake Michigan.

The utter madness of this program can be judged by comparing it with the far less costly system of branch canals that almost impoverished wealthy New York State. Some work was done on most of the Indiana projects but only two or three were ever completed, the only one of importance being the Wabash and Erie, which finally reached Evansville and the Ohio River twenty years later.

East of Fort Wayne no work was being done on the vital connection with the Maumee. Ohio was pushing northward with construction of the Miami and Erie, but it was apparent that building the section westward to the Indiana line would not be undertaken until the Miami reached the valley of the Maumee. Therefore, Williams, with his usual good sense, kept his work force (now numbering two thousand men) together on the Wabash. Peru was passed and before the end of the following year the channel had been dug as far west as Logansport. He needed water and he got it by tapping Deer Creek. Although the Wabash was now halfway across the state, it could not be expected to collect more than a nominal amount of tolls until the way to the Toledo and Lake Erie markets had been opened.

The panic of 1837 had less effect on Indiana than on Ohio and most other states. It felt the decline in the price of farm products, but wheat, corn and the other hard grains could be held back until the market recovered.

Over the years the few hundred Indians remaining in the broad central valley of the Wabash had been coerced into ceding their lands to the state. On their promise to migrate to the open territory west of the Mississippi, they had been paid a dollar an acre for their land. Some had complied and been placed on the Potawatomi Reservation at Council Bluffs, on the Missouri. Those who remained showed no inclination to leave. With land in the vicinity of the canal becoming extremely valuable, local officials took it upon themselves to remove the Indians by force. In all, 859 Potawatomi and Miami men, women and children were rounded up at Grand Prairie. On foot, with armed white men in wagons and on horseback herding them on, the unhappy wretches set out on the long walk to the Illinois line.

It was midsummer. In the oppressive heat old men dropped and were left to die. A number of young children perished. In three days they were marched 120 miles, hungry and helpless. All that was left of their birthright was the name they had bequeathed the state—Indiana. "By the time the pitiful spectacle passed through Terre Haute," Madeline Sadler Waggoner, the Indiana historian, has commented, "most of the children and all of the papooses were dead, their small stiff bodies still strapped to their mothers' backs." [3]

No white man ever faced trial for his part in the Indian removal. The whole matter was brushed aside as rumors began to circulate that the state was teetering on the brink of a financial precipice. In his message to the legislature in December, 1838, Governor David Wallace informed that body that $1,693,000 had been spent on internal improvements. This, of course, was borrowed money, the interest on which amounted to $193,350 while the revenue derived from taxation was only $45,000. "If this does not startle us," he declared, "it should at least awaken us."

21

The Wabash and Erie—
Our Longest Canal

F OR THE FIRST time the fund commissioners came under attack. A small group of dissidents demanded their dismissal. They were shouted down. After the furor died away, it was the consensus of the legislators that the commissioners had erred only in attempting to push forward on too many projects at the same time, and that they be directed to finish one improvement before proceeding with another.

However wise this might have been, it satisfied no one. The only matter on which the state as a whole could agree was that the Wabash and Erie must be completed. But in the meantime, were other improvements, on which thousands of dollars had been expended, to be left to whither and decay? The answer was no, and the supporters of the Whitewater, White River, Eel River and Central canals prevailed. With nothing changed, the Canal Commission continued on its muddled course until it was brought up short by the recession and near panic of 1839 when money suddenly became so tight that it could not borrow additional funds nor pay its bills. In August all work came to a stop.

At first, people could not believe that the state was bankrupt, only to stand aghast as the gross mismanagement and corruption of the fund commissioners were revealed. No books had been kept and no reports rendered other than those that passed between Chairman Milton Stapp and Commission Secretary Dr. O. Coe, the chief culprits. Taking advantage of the wide authority granted them,

Still hoping to find a cargo, this empty boat rides high in the water as it points for Lafayette on the Wabash and Erie Canal.

they had put out on credit no less than $15,000,000 worth of the state's bonds, most of them to the Morris Canal and Banking Company of New Jersey and New York City, which you will recall as the builders of the Morris Canal, a respectable company until it went into the wildcat banking business.

Coe's authority can be measured by the fact that he established his office in the financial district of New York within a block of the Morris Bank, in which he was a director. Coe's method of disposing of the bonds at a profit was ingenious. He made theoretical sales of them to the Morris at a stated figure, then, ostensibly as their agent, disposed of them to other wildcat banks, splitting the profits with the Morris officials. When other irregular transactions, not connected with Coe, forced the Morris Company into bankruptcy, it had no assets and owed the state of Indiana $2,536,611. An audit of its books in 1842 revealed that Stapp and Coe had embezzled more than $2,000,000.

Congress was appealed to for relief but none was forthcoming. By the time the legislature convened in January, 1840, the situation was desperate. Hard money had gone into hiding and the business of the state was at a standstill. In the emergency political factions joined ranks and ordered the sale or abandonment of all internal improvements excepting the Whitewater Canal and the Wabash and Erie. (The Whitewater Canal was later sold to a private corporation which completed it.) To tide the state over the emergency, the legislature authorized the issuance of $1,200,000 in treasury notes or scrip. The paper money was printed in several colors and was promptly nicknamed—and not humorously—Red Dog, White Dog and Blue Dog. At first it was accepted at par, but in two months it depreciated to forty cents on the dollar, which led to the bankruptcy of many country merchants who had received it at face value.

Work on the Wabash and Erie was resumed, but with little being accomplished as the men daily threatened to walk off the job if their wages were not raised to take care of the difference between hard currency and the depreciated scrip they were being forced to accept. The dispute was settled when the contractors gave into their demands.

When the channel reached the Tippecanoe, Williams, still concerned about his water supply, paused to throw a dam across that stream. The delay cost him several months, but by early June, 1841, he put the Wabash and Erie into Lafayette. Water was let into the channel two weeks later and the canal was declared open to navigation from Lafayette to Fort Wayne. From the latter place, in mid-July, the first freighters and packet set out on the long journey across Indiana.

The *Tippecanoe Journal* had informed its readers that the boats were on their way and well in advance had alerted them to the day they might be expected to reach Lafayette. The news had a dynamic effect on the region. For a week wagons from as far as sixty miles away, heavy with grain, began reaching town at the rate of a hundred a day. In many instances sunbonneted wives shared the spring seat with their husbands, anxious as any male to be on hand for the occasion.

"The streets are daily more crowded with wagons loaded with grain," wrote Dr. Timothy Flint, the New England clergyman and author. "Courthouse Square is a confused mass of farmers, waggons, horses and campfires, waiting for the boats to carry their produce away."

If the pioneer freighters and packets had only their practicality to recommend them, that soon changed and the best of the passenger carriers became famous for their elegance, comfort and identity. Most Hoosiers had only a slight acquaintance with luxury but they had a keen appreciation of it, and when traveling by canal, they didn't mind waiting a day or two before setting out so that they could take passage on such elegant packets as the *Indiana* or the *Silver Bell*.

No matter what the canal, its finest packets were always described as "the equal of any ever seen on the Old Erie." That appears to have been true of the *Silver Bell*. Built in Fort Wayne by men whose previous knowledge of shipbuilding was negligible, her lines were graceful and her bow pleasantly rakish. The floor of her main cabin was covered with imported Brussels carpet. Drapes of French lace shielded the windows against glaring sunlight.

Dangling from the ceiling a number of small silver bells tinkled musically with every movement of the boat and vagrant breeze. On the towpath her team of three silver-gray mules in their burnished silver harness informed the countryside that the *Silver Bell* was passing.

Transportation charges via the canal were so much cheaper than by wagon or pack train that the savings were soon reflected in the lowering of retail prices on a a host of items. Further encouragement came when the Congress, aware of the state's distressed financial position, made another grant of public lands. There was a market for them, and sales totaling $175,000 were made. In January, 1842, an audit showed that tolls collected on the Wabash and Erie amounted to more than $22,000. But helpful as these sums were, they barely made a dent in the state's indebtedness. However, there was real hope of future solvency in the rapid growth of Indiana's population and the increased taxes such growth would provide.

What to do about the debt became a burning question. The southern counties demanded that Indiana repudiate its obligations as Michigan and Illinois were threatening to do. The northern half of the state firmly opposed taking such action,

High point of the day—the arrival of the evening packet.

protesting that the debt had been contracted in good faith, and that if Indiana was to preserve its integrity, it must be paid, no matter how long it took.

Unable to budge their opponents, the south surrendered and the legislature got on with its business. Extension of the Wabash and Erie to Terre Haute was authorized.

Williams headed south from Lafayette with a work force of close to one thousand men. From personal observation he was acquainted with the country ahead of him. He had had many setbacks in driving the Wabash and Erie across the state but felt the worst was behind him now. He was to find he was mistaken.

If it had not been for the recurring delays when for lack of funds or due to the ravages of cholera no work could be done, he might have put the canal into Terre Haute in three years. Instead, he was five years getting there. The region south of Lafayette was one of the most thinly populated in the state, averaging not more than seventeen persons to the square mile. There were no towns, only small villages. Country merchants, realizing that the canal contractors and their crews had no choice but to trade with them, raised their prices and gouged them unmercifully.

A pestilence of serious proportions struck the big labor camp in July, felling several hundred men and bringing death to many. Williams, who had seen the same thing happen on the Ohio and Erie Canal, realized at once that it was cholera. Doctors were summoned and they succeeded in checking the outbreak before it reached epidemic proportions. But they did not stamp out the virus, and the laborers on the Wabash and Erie were cursed with repeated sieges of cholera until the canal was completed.

The cholera epidemic was only one of many problems that Williams was to face. The long, dry summer of 1842 convinced him that he could not depend on the Wabash River to supply the canal with the water it was going to need. Dams and feeders would have to be built.

East of Fort Wayne the connection that was to join the Wabash with the Miami Canal at the state line had been completed. The Miami had reached the Maumee River, and Ohio was at work at last on the link that was to join the two canals. Construction was completed early in May, and several weeks later it was announced that on July 4 at Fort Wayne, with ceremonies befitting the great occasion, the Wabash and Erie Canal would be thrown open to navigation for its entire length of 240 miles from Lafayette to Toledo.

The news was received with wild enthusiasm. At last, after years of waiting, Indiana was to escape from its isolation and have direct connection with the flourishing markets of the East—not only with Toledo and the Lake Erie ports but, it was optimistically believed, with New York and the Atlantic seaboard. Now, said the enthusiasts, the canal would begin repaying the cost of building it.

That the opening of the Wabash and Erie would materially affect the economy of northern Indiana was not to be doubted. General Lewis Cass, the principal speaker, took that for his theme in addressing the several thousand Hoosiers who crowded into Fort Wayne for the Fourth of July festivities. The

Heavily laden grain boat on the Wabash and Erie Canal.

general, a hero of the War of 1812 and currently United States minister to France who had been known to charm even Indians with his oratory, had no difficulty convincing his audience that "this great public improvement which we are dedicating today will increase the prosperity of every man in sound of my voice." [1]

The enthusiasm engendered in Fort Wayne was reflected in every village and hamlet across the state, and when, a few weeks later, emigrants began streaming in, crowding every passing freighter and packet, and mills and factories could be seen going up, who could doubt that a new day was dawning for backwoods Indiana?

The guarded optimism with which Jesse Hill set out on his return from the Fort Wayne celebration to the head of construction below Lafayette was slowly dissipated as he observed the decaying condition of the locks and dams he had installed in the past. In a few years they would have to be rebuilt or replaced. He had little reason to hope that the state would find money for making costly repairs without bringing work to a halt on the extension to Terre Haute. That proved to be the case, new construction being disrupted for months on end.

Indiana's great drouth of 1846, in which no rain fell for three months, involved the Wabash and Erie, through no fault of its own, in what local historians dubbed the Attica-Covington or Fountain County War. The canal had reached Covington, the county seat, fourteen miles below Attica. Both villages had been designated canal ports. When the clay-colored Wabash River, lower than anyone remembered having seen it, was let in, the water backed up against the Attica lock and made the canal navigable to that point. Armed with axe handles and ancient fowling pieces, the whole town turned out, determined that the lock would remain closed, thus guaranteeing that for the remainder of the summer Attica would be the southernmost terminus of the Wabash and Erie.

Canal employees were brushed aside and preparations were made against an expected attack from Covington. There had been bad blood between the two towns ever since Covington had captured the county seat by base trickery.

News of what had transpired at the Attica lock reached Covington before evening. A mass meeting was called and officers named for an attack on the Atticans. The following morning a force of two hundred men, some on foot and others in wagons or on horseback, set out for the north, determined to open the lock and prepared to burn powder if necessary.

The outnumbered Atticans put up a brief resistance. There was some shooting and a few heads got cracked, but no one was killed or seriously maimed. The lock gates were opened and put out of commission. The impounded water rushed down the dry canal bed, only to disappear before it reached Covington. After the invaders withdrew and all attempts to close the gates failed, tons of hay were dumped into the opening, but the yellow water seeped through. The real losers were the boats that had been caught in the lock. As the water fell, they sank into the mud and remained there for weeks.

The Wabash and Erie put its first boats into Terre Haute in 1849. To get there it had had to overcome one problem after another: work stoppages, floods, drouth, epidemics that shattered its work force and the ever-present threat that funds would be cut off.

Terre Haute was the jumping-off place for the Illinois country and the Far West. Its population doubled within two years after the arrival of the canal. No

Impoverished Indiana's first capitol building at Vincennes.

one was more surprised than Jesse Williams when the state suddenly ordered the Wabash and Erie pushed through to Evansville and the Ohio River. Aside from the old French town of Vincennes, southwestern Indiana was still largely an un-tracked wilderness. The much-publicized National Road that was supposed to open up that country was still only an impassable stump-filled trail.

The least costly route for the canal to take was to dig the short cut that had been proposed in the Internal Improvements Bill of 1836 but on which no work had been done; then down Eel River to the west fork of White River, which could be followed to its conjunction with the east fork, from where sixty-six miles of channel would have to be dug due south to Evansville. A reservoir and dam

Log tavern on the banks of the Wabash and Erie Canal at Huntington.

at Birch Creek would provide the canal with adequate water. This was the route Williams elected to take, after some lines had been run.

Bad luck continued to pursue him. Construction of the Birch Creek dam had just begun when another outbreak of cholera brought operations to a halt. Months were lost before work could be resumed. With one delay after another, it was not until June, 1852, that the first boats reached Mayville, 392 miles from Toledo. Completion of the great canal was in sight when the Birch Creek reservoir and dam were destroyed by bombs. There wasn't much question but that the farmers in the surrounding territory were responsible for the destruction. The state sent in marshals and militia, but the identity of the perpetrators was not discovered. It was a serious setback, leaving Williams no choice but to return to Birch Creek and restore the dam. As the work was being finished a second attempt to destroy it was aborted by the vigilance of the guards.

Although the violence that erupted at Birch Creek was not repeated else-where, it was becoming obvious that a widespread disenchantment with canals was taking place. More and more people were becoming convinced that canal reservoirs, with their stagnant water, were somehow responsible for the constant outbreaks of cholera. But far more persuasive were the steadily rising taxes that the state was levying to whittle down its great canal debt. On the other hand, there were the railroads, half a dozen of which were in operation (though none of any great length), not costing Indiana a penny and actually paying their share of the tax burden.

With little of the pomp and ceremony that once would have attended the event, the Wabash and Erie Canal was completed to Evansville in 1856—458 miles, the longest canal ever dug in the United States.

But for only four years was it to remain open to navigation from end to end, and they were to be years of disaster piled on disaster: locks destroyed and channel washed away by floods. In 1857 its tolls amounted to only $60,000, while repairs below Terre Haute alone had cost $40,000. In January, 1858, the legislature ordered that any section of the Wabash and Erie not meeting expenses be closed. Several attempts to keep the canal open below Terre Haute were made, but in 1860 it was abandoned forever. Tolls were steadily decreasing on the northern section. The Chicago, New Albany and Louisville Railroad, originally the New Albany and Salem,[2] had reached Michigan City, providing what it termed "fast through service from the Ohio River to Lake Michigan." Inevitably it drained business away from the canal, while from the east a far more formidable competitor was bearing down on the old waterway, namely, the Lake Erie, Wabash and St. Louis Railroad (the beginning of the great Wabash System). When its rails reached Logansport, halfway across the state, it took dead aim on the canal and practically put it out of business by lowering rates.

The Wabash and Erie struggled on for another two years. On October 26, 1872, the last boat through for Toledo left Terre Haute. Efforts were made to keep short sections of the canal open but in 1874 the state closed it forever.

Bedeviled almost from the beginning by debt, corruption of its officials and legislative folly, the Wabash and Erie Canal was a financial failure, but its great contribution to the opening up and advancement of Indiana could not be measured in dollars.

Its demise was dramatic. Several hundred rustics had gathered at the Deer Creek Aqueduct to witness the passing of the last boat on the canal east of Lafayette. They cried out in horror as the old structure gave way and mules and Negro driver were swept down into the creek and drowned in the roaring water.

Perhaps it was a fitting end for the Wabash and Erie Canal.

22

The Louisville and Portland— the Little Bonanza

THE FALLS OF THE OHIO, at Louisville on the Kentucky shore and opposite Jeffersonville on the Indiana bank, could not have been much of a hindrance to the Indian paddling his canoe up and down the great river; nor could the early white pioneers who poled their arks and rafts downstream have found them more than a nuisance. To the steamboaters, however, they were a menace to navigation.

When talk developed of building a short canal to bypass the falls, Cincinnati, which had visions of becoming a great river port, supported the idea. Ohio, Pennsylvania, Virginia and Kentucky appointed a commission to look into the feasibility of such a program and estimate its probable cost. The commissioners spent several years deliberating the matter. In the meantime, the new state of Indiana, which had not been invited to take part in the proceedings, incorporated the Ohio Canal Company and began some preliminary work on a channel. This spurred the commission into action. Meeting at Pittsburgh, it rendered a lengthy report recommending construction of a short canal which, if dug on the Kentucky shore of the Ohio, should not cost more than $400,000, or less than half the sum that would be required to install it on the Indiana side of the river.

That silenced Indiana. Years passed before anything further was done. In 1825 the Kentucky legislature chartered the Louisville and Portland Canal Company, which was empowered to construct a 2½-mile-long canal on the Kentucky shore from a point above Louisville to the settlement below the falls, which appropriately was known as Shippingport. The company was incorporated for $500,000. The federal government subscribed for $250,000, private capitalists underwriting the balance.

Violent opposition to the undertaking erupted at once. Understandably, much of it came from that segment of the population whose livelihood would be swept away by the improvement, namely, the hundreds of men who were engaged in carting freight between the two points where the steamboats were discharging it. A canal would make the transshipment of freight unnecessary.

Just as firmly opposed to the idea were the majority of Louisville's merchants and professional men, who protested that a canal that bypassed the falls would be the ruin of the town. Up to now Louisville had been the accepted stopping place for all river traffic; build a canal and the boats would stream past Louisville without spending a dollar for food, fuel or other necessaries.

Although the stock of the Louisville and Portland Canal Company was fully subscribed, money was slow coming in and construction did not get under way until 1828. It was a year of extremely high water on the Ohio. Before the work could make much progress, the summer was half gone. The hostility of many property owners resulted in endless condemnation proceedings before a right of way across the Louisville waterfront could be secured.

Despite the delays and harassment, digging the channel was accomplished before winter set in. All that remained to be done was to smooth off the bottom of the cut and pave the sides with rock. The rock work was let out in sections

Steamboat entering the short Louisville and Portland Canal. *Courtesy Library of Congress.*

to subcontractors. Henry A. Jones of Cincinnati, then a youth of twenty-one, later to become famous as an Ohio River steamboat captain and owner, was awarded contracts for building two sections of the work. His only previous employment had been as a laborer on the Ohio and Erie Canal for two years. Reminiscing in later years, he saw nothing incongruous in his having become a contractor on the Louisville and Portland at twenty-one. "I possessed the necessary articles for the work to be done, viz: a willingness to work, a wheelbarrow, a hammer and a strong arm."

The first steamboats passed through the Louisville and Portland Canal in 1830. It took just one full season of operation to convince the disgruntled citizens of Louisville that the channel was a godsend. Louisville became the busiest port city between Pittsburgh, St. Louis and New Orleans. Hundreds of the smaller, shallow draft boats made it their terminus, discharged their miscellaneous cargo there, picked up freight the big boats had left and returned up the Ohio to points the great sidewheelers could not reach.

Contrary to predictions, Louisville grew, and so did Shippingport, until they finally merged. It was a time when every town and hamlet on the Ohio and its tributaries was turning out boats of one kind or another; across the river from Louisville, at New Albany and Jeffersonville, the Howards and other yards were building such famous luxury packets as the *Eclipse* and the *Grand Republic,* with half a hundred other floating palaces to follow.

Consider these official figures of the number of boats passing through the Louisville and Portland Canal in 1831: 416 steamboats, 46 keelboats and 357 flatboats, totaling 76,000 tons. Compute this against tolls of $.20 per ton on steamboats and $4.00 on rafts and flatboats, which the company was authorized

to collect, and it will be seen that even in its first year of operation the Louisville and Portland was a bonanza. It had cost less than $750,000 to build, only four locks being required to accomplish the sixty-eight-foot drop from above Louisville to Shippingport.

Under the terms of its charter the directors of the canal company could raise its tolls whenever they failed to produce a profit of 12½ percent, the legislature reserving the right to reduce them if and when they exceeded 18 percent. Had the legislature exercised this right steamboat owners would have had little reason to complain. But it didn't. Without the state taking any action, tolls were raised repeatedly.

By 1835 the Louisville and Portland annual earnings were proven to be in excess of 25 percent. It became a public scandal with several political leaders close to Acting Governor Morehead being accused of corruption. The storm blew over, however, without having changed anything and the Louisville and Portland continued to reward its stockholders with handsome dividends. After the Old Erie, it was the most profitable of American canals. The company had made only one mistake: they had built their canal too small.

Steamboats of ever-increasing size had been coming off the ways since the mid-thirties. When the first of the three *J. M. White*s was launched, she was the biggest and finest steamboat on America's inland waters.[1] Five years later she was considered "small" in comparison to the magnificent three-deckers that were plying the Ohio and Mississippi. Looking ahead, as the big boats became more numerous, it was apparent that the situation that had led to the building of the Louisville and Portland Canal was about to be repeated.

The crunch came when the *Jacob Strader* of Pittsburgh, bound for New Orleans on her maiden voyage, found that she could not squeeze through the canal. Passengers and cargo had to be transferred to a smaller boat. It was the first of many such frustrating and costly incidents. Wealthy Captain W. J. Kountz, of Allegheny, Pennsylvania, president of the well-known Kountz Lines and dean of Ohio River steamboatmen, lost no time taking advantage of the situation.[2] Within ninety days he put in the water the first of what he termed his "short boats." As the name implied, they were short—125 feet overall—and slow, of shallow draft, but able to handle a deck load of a hundred tons. It wasn't long before a dozen or more of these ugly ducklings were steaming up and down the Louisville and Portland, transporting freight from the mouth of the canal to Shippingport.

In 1855 the federal government, exercising its still-disputed authority for regulating interstate commerce, ordered the enlargement of the Louisville and Portland, and in a supervisory capacity took an active part in the work. The Civil War intervened and half a dozen years were lost. When the enlargement was finally completed in 1872, it passed out of private ownership and came under control of the Congress. Competently administered it has served for years as an important link in our complex transportation system.

In the meantime, due to the unremitting labors of the Corps of U.S. Army Engineers, the falls of the Ohio are no longer a menace to navigation.

23

The Illinois and Michigan—
Joliet's Canal

THE STORY OF AMERICAN CANALS can be said to have begun on that evening back in 1673, when Louis Joliet, the French explorer, and his companion, Father Marquette, bivouacked on the short portage between the little Des Plaines River and the sluggish stream that the Indians called Checagou (Chicago). "M. Joliet expounded at great length on the benefits that would be derived from digging a canal to connect the two waterways," wrote Marquette, "permitting passage from Lake Michigan down the Illinois River to the Mississippi."

The matter was resolved on July 4, 1836, when after ten years of bickering and false starts, the infant state of Illinois broke ground for the Illinois and Michigan Canal.

Although its land area of 58,329 square miles was 40 percent larger than neighboring Indiana's, Illinois had less than half as many inhabitants when it was granted statehood in 1818. North of the old French settlements around Cahokia, the only villages of any size were Peoria and Galena, the mining town on the Mississippi. A hundred or more settlers scattered about on the bogs and sloughs in the vicinity of Fort Dearborn formed the nucleus of what was to become Chicago. Indians outnumbered whites four to one.

In 1803 the United States built Fort Dearborn on the south fork of the main channel of the Chicago River and stationed a company of troops there to control

Locks, Illinois and Michigan Canal. From a painting by John D. McKee.

the Indians. Although Fort Dearborn was little more than a stockade that could not be defended against attack, it sufficed until British agents during the War of 1812 aroused the Indians and they burned Fort Dearborn and massacred twenty-six soldiers, twelve trappers, two women and twelve children. Troops were rushed in and the Indians, mostly Potawatomi, were severely mauled for their bloody insurrection.

Fort Dearborn was rebuilt in 1816. By then the great stream of emigration from the East had begun, as many as a hundred wagons a day arriving at The Forks (Chicago) and either stopping there or continuing on westward for another hundred miles, where they had their choice of millions of acres of prairie land that was being offered by the federal government at the reduced price of $1.25 an acre.

From the lake came other hundreds of movers who had driven across Michigan to St. Joseph and taken passage there by schooner for Chicago, where with some difficulty they were landed. That obstacle was not removed until 1834, when two five-hundred-foot piers were built, creating a new channel for the river, and the first schooner, the *Illinois*, was able to sail into the harbor.

Far to the south in that part of the state bordering the Ohio, hundreds of Kentuckians were crossing the river and settling in Illinois. Almost without exception they were the sons or grandsons of mountain men who had tamed the wilderness of Kentucky and they brought with them their own rules of conduct and conception of justice. Many of them were or had been slave owners. In their attitude about slavery they were bringing into Illinois the seeds of a conflict that was to divide the nation as well as the state less than half a century later.

Illinois wasn't all undulating, treeless prairie that required only the turning of the sod to put it under cultivation. Most of the hilly country south of today's Shelby County to the confluence of the Ohio and Mississippi at Cairo was heavily wooded. Great fields of bituminous coal lay close to the surface in many places, especially in Jackson and Williamson counties. But no market existed for coal. On the other hand, sawmills were clamoring for timber. There was so little law in the country that organized gangs of timber thieves operated almost at will. A man who owned a valuable stand of hardwood was apt to find after an absence of several days from his property that his grove had disappeared. Vigilante groups were formed, and there followed the usual shootings and unexplained killings.

Whether John Kinzie, the Indian trader and real founder of Chicago, was acquainted with Louis Joliet's speculations about a canal to connect the Illinois River with Lake Michigan is unknown. But it is certain Nathaniel Pope, territorial delegate from Illinois, was. It was he who convinced his friend President James Madison that such a waterway would be of inestimable value to the United States, that a canal connecting the lake with the Illinois by means of the Chicago River and the Des Plaines would open a trade route to St. Louis, the Mississippi River, New Orleans and salt water. It must have aroused Madison's interest. In 1814 he asked Congress to authorize the construction of such a waterway.

His request was brushed aside as visionary. However, the project which had won the support of Secretary of the Treasury Gallatin and Peter B. Porter of the New York delegation, a canal enthusiast, was kept alive in Congress for thirteen years. Even more persistent and persuasive was Illinois Representative Daniel P. Cook, for whom Cook County was named. Undoubtedly Stephen A. Douglas, the Little Giant, and Abe Lincoln were the greatest stump speakers Illinois ever produced. But Dan Cook was nearly in their class. Largely to silence him, Congress in 1822 granted a right of way across public lands for a canal but did not give the state any land to be sold in aid of the project. Actually the so-called public lands were in part Indian lands under existing treaties; the government overcame that difficulty by getting a number of Potawatomi chiefs to affix their mark to a paper approving the grant.

Farmers of north-central Illinois, who could find no profitable way of disposing of the bountiful crops they were producing, were almost unanimous in demanding that the state build a canal that would give them access to a market. They knew that Illinois did not have the money with which to undertake such a costly improvement, but that did not silence them. Borrow it, they argued; a canal would pay for itself in ten years.

"The Legislature was under pressure from other sources," wrote historian A. T. Andreas.

Steamboats were now operating on Lake Michigan and in the course of a season as many as 175 schooners were reaching the port of Chicago. Two years before, it had been considered remarkable to see a hundred wagons arriving in a day; they were now coming in twice that number. The roads from here [Chicago] to La Salle and Peoria were thickly dotted with wagons.[1]

Shallow-draft steamboats were ascending the Illinois River to Peoria, and at favorable times of the year continuing on to La Salle, the western terminus of the proposed canal. It was widely believed that by deepening the channel of the Chicago River at its mouth the lake would flow into it and find its own way at ground level down the Des Plaines to the Illinois.

To settle the question, the state appointed a board of Canal Commissioners in 1825 and they hired a surveyor named Bucklin to stake out a route. He made no claim to being an engineer, but he knew how to run levels. His report came as a surprise. According to the figures he submitted, Lake Michigan was about 150 feet higher than the point where the canal would reach La Salle; and while lake water would not meet the river in a direct descent, the summit of the intervening watershed was not more than five feet. He estimated the cost of constructing the canal at $700,000.

This was good news, being less than half of what the commissioners had expected the cost to be. Illinois appealed to the federal government for aid in building the canal. When a year passed without the Congress taking any action, the state in January, 1825, granted a charter to a private corporation, capitalized (theoretically) at a million dollars, to build the Illinois and Michigan Canal. Turning over such a potentially rich source of revenue to a corporation was considered an outrage by many voters. Their anger cooled a year later when the Illinois and Michigan Canal Company, unable to dispose of its stock, surrendered its charter. They were further rewarded in 1825, the Congress having finally got around to granting Illinois 300,000 acres of land lying along the route of the proposed canal.

The canal commissioners prepared new plans and platted the towns of Chicago and Ottawa, offering hundred-foot lots and building sites for sale. A disappointing amount of business was done (this was prior to the years of wild speculation). A bad storm swept down the lake putting the waterfront and all of what is today the Loop under a foot of water and leaving it a soggy mess. Nevertheless Chicago was made an incorporated town of the second class. The first work done by its municipal officials was construction of a wooden bridge over the Chicago River at today's Dearborn Street.

Meanwhile the Canal Commission, unable to raise or borrow money, could not proceed. Worse still, it had become obvious that the estimates rendered by Bucklin were worthless, that the cost of building a water-level canal would run into the millions. The commissioners tendered their resignations and reported their conclusion that a railroad would serve the public better than a canal, would be cheaper to build and could be kept in operation the year round. The report was no sooner made public, along with Governor Reynolds' approval, than the state found itself divided between pro-canal and pro-railroad factions.

The federal government, which had been slow to respond to the needs of Illinois, was now engaged in proceedings to stifle the claims of Indians to their lands and was removing them from the state. After months of negotiations, eight thousand Chippewa, Ottawa and Potawatomi gathered on the outskirts of Chicago and their chiefs signed or put their mark on a treaty with the United States commissioners by which they relinquished all claims to their lands east of the Mississippi, amounting to five million acres. In return the government agreed to pay the Indians $1,000,000 in money and supplies over a period of twenty-five years, and to place them within a period of two years on an area of equal size in Kansas, Missouri or Iowa.

A down payment of $150,000, mostly in goods, was made. Wrote Elias Colbert:

> It is reported that not less than twenty thousand dollars' worth of the goods were stolen by the Indian traders during the first two nights, after the owners had been liberally saturated with whisky, for which they paid out a large proportion of the articles furnished. A letter from a traveller who witnessed the scene, was unearthed and published in the *Tribune* in 1869. The description there given of the disgusting revels of the red men, and the rapacity of the whites, was almost enough to make one lose faith in human nature.[2]

Another payment of $30,000 in goods was made to four thousand Indians in October, 1834.

"The scene was simply disgusting," says Colbert. "Several of the Indians were killed in a drunken brawl."

The Indians were reluctant to leave their ancestral homeland and employed one excuse after another for delaying their departure. Finally, the government's patience was exhausted; troops moved in the following August and started the Indians on their long trek westward, some in wagons, many on foot. At East Dubuque they were ferried across the Mississippi and dispersed to various reservations.

Chicago had had an amazing growth in the past ten months. The weekly *Democrat* boasted that the town now had a population of twelve hundred. "The hotels and boarding houses were always full," says Andreas:

> and full meant three in a bed, with the floor covered besides. Many of the emigrants lived in their own covered wagons, others in a rude camp, hastily built, for home or shelter. All about the outskirts of the settlement was a cordon of prairie schooners, with tethered horses in between, interspersed with camp fires at which busy housewives were ever preparing meals for the voracious pioneers.

In the same vein, the *Democrat* said, "Hardly a vessel arrives that is not crowded with emigrants, and the stage that now runs twice a week from the East is thronged with travelers. The steamship *Pioneer*, which now performs her regular trips to St. Joseph, is also a great accommodation to the traveling community. Loaded teams and covered wagons, laden with families and goods, are daily arriving and settling upon the country back."

Both the *Democrat* and the *Tribune* chortled over the fact that the schooner *Eliza Little*, from Chicago, reached Buffalo in the spring of 1835, two weeks before the first boat of the year came through the Erie Canal. They noted also

The Illinois and Michigan Canal connecting Lake Michigan with the upper Mississippi.

that steamboats on the Illinois River were reaching Peoria regularly and in stages of high water proceeding an additional forty-five miles to La Salle, less than fifty miles from Chicago.

Although construction of the Illinois and Michigan Canal appeared to be as far off as ever, whether to dig or not to dig remained the most hotly debated question throughout the state. If you were anti-canal, it followed automatically that you were pro-railroad. The argument put forth by that faction ignored the fact that a rail line, no matter how long (they envisioned one extending down through the state from the Wisconsin boundary to the vicinity of Cairo), would have no road with which it could connect, for as yet not a foot of track had been laid in any neighboring state.

As election time neared, it became obvious that ballots would be cast for a candidate solely because he was for or against the Lake Michigan and Illinois River Canal. Joseph Duncan, the Democratic candidate for governor, a staunch canal advocate, was elected with votes to spare. He accepted his victory as a "mandate" (a term that has worn threadbare in American politics) to begin construction of the canal without further delay. Soon after being inaugurated he named a new commission with power to raise funds, engage a competent engineer and begin work. Sales of land granted by the Congress not having produced enough money to get the project off to a good start, an emissary was sent East to negotiate a loan of $500,000 on state bonds. As we have seen, this was the accepted practice of needy states. The loan was secured, and on July 4, 1836, with the usual ceremonies, ground was broken at Canalport (near Joliet) and the digging got under way. Among those on hand for the occasion was a self-effacing young man from western New York State by the name of William Butler Ogden, who was to become Chicago's First Citizen and greatest benefactor.[3]

The wave of speculation that was beginning to sweep the country reached Chicago. The first of the professional get-rich-quick operators had arrived and were turning the town upside down. There was no canal as yet and wouldn't be for another twelve years, but along the line of stakes indicating where it would course one day, so-called canal lots had leaped in value. In this region of bogs covered with skunk cabbage and wild onion, the home of colonies of skunks, lots rated worthless a few months back were suddenly bringing $200 to $500 for hundred-foot lots.

The appearance of Chicago was changing. No more log cabins were being built. In their place, houses and stores of original Chicago architecture were being run up. They were ugly but could be quickly put together and were a vast improvement on the log cabin. Heavy posts were sunk into the ground at the corners of the building and in between them slabs of wood, fresh from a sawmill, were laid. On the slabs a foundation was placed to which was spiked a sill; three-by-four scantlings in an upright position were nailed to the sill. The rest of the operation was much the same as builders use today. Because of the slab wood used, the buildings were called "slab houses," and they gave Chicago its nickname of Slab Town. The Great Fire of 1871 destroyed most of them. The rebuilding ushered in the new age of brick, stone and concrete.

The Illinois Canal Commission was more fortunate than it realized when it engaged Henry Gooding, another Erie Canal alumnus, as its chief engineer. Gooding stayed with the I. & M. (the abbreviation was coming into general use) for thirteen years, something of a record in canal work, serving under three governors in that time. When Governor Duncan took office his concept of what the waterway should be was a ship canal through which steamboats could pass from river to lake. Such a prospect delighted Gooding and he submitted plans calling for a channel sixty feet wide at the water surface, thirty-six feet at the bottom, with a six-foot depth. He also presented the commissioners and Governor Duncan with profiles for both a lake-level canal, cut deep enough to permit Lake Michigan to feed it, and a less costly summit-level canal. He estimated that at the current price of supplies and labor the preferable lake-level channel could be built for $4,000,000; the summit-level for $1,750,000.

For a state whose treasury was bare, $4,000,000 was an awesome figure to contemplate. But there was a contagion of prosperity in the air. Turning their backs on the thriftier course they might have pursued, Governor Duncan and his advisors took the plunge and began construction of the lake-level Illinois and Michigan Canal. With its six locks it was to become one of the most costly of all American artificial waterways.

Soon after the work got under way, the "neglected" districts throughout the state insisted, with some justification, that the canal would not benefit them; they banded together and demanded that the state build a system of railroads. In an attempt to placate them, Illinois stepped into a hole that had no bottom. Before she could pull herself out, she was committed to building a thirteen-hundred-mile rail network, a colossal folly that was to bury her under a mountain of debt and end in bankruptcy.

Two thousand men were soon at work on the canal, and in half a hundred places throughout the state small gangs of a hundred or more laborers were making roadbed. Almost without exception they were Irish and Germans, brought in by contractors who had recruited them in New York. Practically all had entered Illinois by way of Chicago and spent a few days there, adding to the excitement that gripped the town as farm and real estate speculation soared to new heights.

An unofficial census conducted by the *Democrat* placed the more or less permanent population at two thousand. This did not take into account the several hundred settlers who reached Chicago and continued on west after spending a day or two replenishing supplies. The business section was beginning to center on Lake and Water streets between State and Franklin. Mud was still everywhere, and many merchants were laying plank sidewalk in front of their stores. But no uniform level having been established, progressing from one street to another was very much like going up and down stairs.

The three-story Lake House, "the finest hotel on the Northwestern frontier," opened its doors in 1836, boasting that its meals were prepared by a French chef. As proof of its elegance it provided its patrons with a printed menu, the first seen in Chicago. Popular Francis Sherman, the prosperous boarding-house keeper, began construction of the first of the series of Sherman Houses that were to distinguish the city for a century and more. He gave the nation the hotel coffee shop, in which the same food served in the main dining room could be had at reduced prices.[4]

With prosperity came inflation; suddenly everything was costing more. The contractors digging the canal could no longer provide board at the stipulated $2 a week. When they raised the price, the men struck for higher wages. The state borrowed another $500,000 and came to the rescue. Lulled to complacency by the hope that the canal and some of the railroads it was sponsoring would soon be finished and money would be rolling in, the governor and his associates saw no cause for alarm. But five months later the boom was over and the nationwide panic of 1837 brought all business to a standstill.

By act of the legislature Chicago became an incorporated city of 4,170 on March 4, empowered to elect its own officials. A citywide election took place on May 2 and William B. Ogden was chosen mayor by a two-to-one margin over his nearest opponent, John H. Kinzie, son of the "Father of Chicago." Like Uncle Dan'l Drew, the make-believe rustic who helped to wreck the Erie Railroad, who once said, "I got to be a millionaire almost before I knowed it," Ogden had become Chicago's richest man and leading real estate operator in two short years. He was not yet a millionaire but was on the way to becoming one. When the state of Illinois went bankrupt and suspended payment of its debts, Ogden, himself on the verge of financial disaster, kept Chicago from doing the same. "No!" he cried, "do not tarnish the honor of our infant city by repudiating our just debts."

Now more than ever it was felt that the Illinois and Michigan Canal must be completed, that it alone could save the state and Chicago. Ogden was called in and entrusted with finding enough money to keep the work going. Somehow it managed to limp along for four years. The state bank failed in 1841. Illinois owed $14,000,000 and interest was piling up at the rate of $830,000 a year. The Canal Commission issued $400,000 in scrip but it quickly depreciated and work on the canal halted.

The state was so discredited that bills against it were sold for as little as eighteen to twenty-four cents on the dollar. In Chicago, lots that had brought $750 to $1,000 a few months back could not be given away.

With Ogden preaching economy and trying to salvage what he could of the grandiose program of internal improvements, Duncan and the commissioners agreed that before further work was done on the canal, the cheaper summit-level course, which they had rejected, should be reconsidered. Gooding was instructed

Old wooden gates at Channahon on the Illinois and Michigan Canal.

Junction of earth and rock cut on Illinois and Michigan Canal.

to render estimates of the cost of completing the I. & M. as planned, against the savings that could be made by switching to the alternate course.

Gooding's figures were startling. If the deep-cut route were followed the cost would be more than $3,000,000; by the other route, $1,600,000. That left no room for argument; the lake-level plan was abandoned and the shallow summit-level course instituted. But plans were only plans, and without funds nothing could be done. For three years Illinois and Chicago—they were almost inseparable—suffered through a period of complete stagnation. The population of Chicago increased by only two hundred.

With the election of Governor Tom Ford in 1842 business began to recover. It was not until 1845, however, that the state was able to borrow enough money to resume work on the canal. Illinois was by now divesting itself of many of the poorly conceived and unprofitable railroads, most of them just twin streaks of rust meandering across the prairie. The decrepit Northern Cross Railroad, fifty miles of junk between the village of Meredosia and Springfield, will do for an example. Its rolling stock consisted of two secondhand locomotives, a combination passenger coach and half a dozen freight cars. Somehow the state had managed to pour $406,233 into it. It was knocked down at auction to a group of Springfield businessmen for $20,000. Of far greater import, and amid cries of graft and corruption, the state granted a charter to a private corporation, known as the Illinois Central Railroad Company, and thereby relieved itself of its greatest burden. So grateful was it to have the Illinois Central taken off its back that it made the Illinois Central a land grant of 2,595,000 acres, which was almost as much as what the railroad had received from the federal government. In return the I. C. railroad agreed to pay the state 7 percent of its gross earnings. It further agreed to build some seven hundred miles of road in six years. It appeared to many that both parties to this deal had lost their wits. The cynics sat back and waited expectantly for an explosion that would rock the state and send the perpetrators of this incredible business to prison for the rest of their natural lives. It never happened. Instead, with the passing years, this proved to be the best bargain Illinois ever made, and the Illinois Central Railroad was to become one of the greatest rail carriers in the United States.

As business recovered, Illinois, the "deadbeat state," began to pay off its long overdue obligations. The Illinois and Michigan Canal was nearing completion. On April 19, 1848, the first boat from La Salle locked through the canal and in midafternoon was floating on Lake Michigan. Chicago, now a city of twenty thousand, hailed the event with a wild mass celebration that lasted for several days. The cheering throngs had more reason to celebrate than they realized. Wheat brought by the I. & M. Canal was to make the town a great grain market, later, with the aid of the Illinois Central, the greatest in the United States. The thousands of hogs and cattle arriving by canal and railroad were to make Chicago the capital of the country's packing industry.

For two decades and more the tonnage handled by the I. & M. taxed its capacity. In 1865 and 1867, its peak years, tolls amounted to $300,000 annually. Official figures place the cost of the canal at $6,557,681, not including interest. Its earnings fell short of that amount by more than $2,000,000, but that was a trifling price to pay for the incalculable benefits it bestowed on Illinois and Chicago. For six years after beginning operation, the I & M.'s passenger business was very profitable, boats leaving Chicago and La Salle morning, noon, and

Fox River aqueduct on Illinois and Michigan Canal.

night on the hundred-mile run to the opposite ends of the canal; but with the opening of the Rock Island Railroad in 1854, which paralleled the canal most of the way, the packet trade quickly became insignificant.[5]

The population of Chicago had soared to 74,500. As yet the city's so-called sewers were open affairs built to drain into the nearest water, which in this case was the river or the lake. The daily death rate from typhoid was extremely high, higher than the combined rate from all other causes. Finally officials got around to asking why. They knew nothing about ecology—doubtless had never heard the word. But they quickly discovered that Chicago had polluted its water supply by dumping sewage into Lake Michigan. At once scare headlines alerted the city to the danger it faced. Other civic problems were forgotten as rich and poor united in demanding that steps be taken at once to safeguard the waters of Lake Michigan.

The startling idea of using the I. & M. Canal to carry away the city's soluble waste and dump it into the Illinois River was advanced and widely supported. The state agreed to the arrangement, provided Chicago shoulder the expense of altering the canal.

Work began at once and by 1871, at an expense of $3,000,000, the canal had been reduced to the lake level and Lake Michigan water was sweeping away from Chicago and down upon the countryside of the Illinois River Valley an endless stream of sludge, garbage, animal and human excrement, befouling the air and turning the once clear and placid canal into a poisonous latrine. To hurry the scum along, the big pumping station at the summit level had been reactivated. The wonder was that boats continued to ply the canal. Tolls were lowered, but the business declined.

The great conflagration of October 8 and 9, 1871, that leveled Chicago was interpreted by some people as an expression of heavenly punishment for the callous manner in which the city had unloosed its filth on its neighbors. However fanciful that idea, the legislature decreed that in view of the great losses the city had sustained it would not be held accountable for the $3,000,000 it had spent in altering the I. & M. Canal.

That deepening the old canal was not going to solve growing Chicago's drainage problem became obvious within a few years. By 1881 the channel had collected so much filth from the city's sewage that it had become putrid, giving off a stench that was intolerable to anyone within miles of it. Something had to be done. To meet a situation that was becoming desperate the commission took some heroic steps. To assure the city clean water it moved the intake cribs a mile farther out into Lake Michigan and at the same time began dredging a new channel from the mouth of the Chicago River to the Illinois, and down that river to its confluence with the Mississippi at Grafton, deep and wide enough to permit the passage of steamboats.

This, of course, was to become the dual-purpose—navigation and sanitary—Chicago Drainage Canal. Along with other improvements, a power plant was built at Lockport, north of Joliet. In its reports the Sanitary Commission stressed that "purification plants which will disperse soluble sewage" would be installed. This last was largely wishful thinking, for in France, where such plants were being tested, they had yet to prove their worth.

It soon became obvious that the Chicago Drainage Canal was to be a major undertaking. Of all the canals built in America only the short-deepwater Chesapeake and Delaware and the New York Barge Canal were more costly. But now Chicago could afford it. In 1890 it was the third largest city in the United States and had drawn unto itself such giants of finance and industry as William Butler Ogden, now president of the Union Pacific and the Chicago and Northwestern, Judge Elbert Gary, the steel magnate, Cyrus McCormick, inventor of the reaper, Philip Armour, Nelson Morris, Gustavus Swift, Montgomery Ward, Joseph Leiter, the grain king, Marshall Field, Potter Palmer, George Pullman and a score more.

Work on the Drainage Canal began in 1892 and on January 2, 1900, the first boat passed through from Lake Michigan to the Illinois River. As a sanitary project the Chicago Drainage Canal accomplished the purpose for which it was built; as a commercial waterway its importance was dwarfed by the network of railroads that soon converged on Chicago, making it the undisputed transportation capital of America.

Almost unnoticed the Illinois and Michigan Canal was abandoned and the old channel left to fill up and disappear.

24

The Lotteries and Corruption

EVERY DAY IN THE ghettos of American cities millions of men and women briefly escape from the hopelessness of their situation by wagering their nickels and dimes against the terrific odds of the "numbers" or "policy" game. If they win, and enough do to keep the racket flourishing, the returns are high. If they lose today, there is some comfort in knowing that there is always tomorrow.

The players know little or nothing about the inner workings of the "syndicate" that controls the game. Maybe it is operated by organized crime. They couldn't care less. Since a man can be put in jail for buying, selling or being caught with a "policy" slip in his possession, they are a party to an unlawful business which annually outgrosses every other gambling operation in the United States, making the numerous lotteries that flourished in canal days seem inconsequential. Throughout the nineteenth century, however, Americans spent more money on lottery tickets than on all other games of chance combined.

The lottery, whether conducted in behalf of some municipal improvement or for governmental or private gain, was patterned after the Italian *lotteria*. In the United States, in the beginning, raising funds for any worthwhile cause by means of a lottery was considered to be a highly respectable business. The clergy, college trustees and the heads of patriotic societies employed it as a painless way of securing money for a new bell, church roof, dormitory or statue to be placed on the village green.

During the years when more than fifty canals were under construction or being projected in that part of the United States east of the Appalachians, all or nearly all were financed at least in part by money raised through lotteries. The great exception was the Erie and the branch canals of New York, because the state refused to legalize that form of gambling. The irony, of course, is that today New York conducts its own lottery with prizes totaling $1,000,000 or more, for which tickets can be purchased at any bank, gas station or practically any establishment that has a roof over it.[1]

Between 1825 and 1845, when Pennsylvania was suffering a sustained attack of canal madness, it granted a lottery privilege to more than fifty different concerns. Many of these manipulators were so dishonest that their scheming produced a public scandal of such proportions that Pennsylvania joined New York in outlawing all lotteries. It has been estimated that overall $51,000,000 raised by means of lotteries actually was spent on canal construction. How many additional millions the swindlers garnered from a gullible public with their rigged lotteries and worthless stock in canals that were begun without any intention of being finished will never be known.

When a group of respected and financially responsible men had organized themselves into a canal company and received a charter from the state through which their waterway was to run, they seldom had any difficulty in securing permission to conduct a lottery for the purpose of raising funds with which to proceed with construction. At this point they invariably called in a firm of professional promoters to conduct and manage their lottery, giving them in most instances authority to proceed as they pleased.

This arrangement resulted in the lottery firms skimming off a major share of the money that was raised and the canal company receiving little or nothing. But before getting into that, a word about America's most famous, richest and longest-lived nineteenth-century gambling extravaganza, the Louisiana State Lottery. Its tickets were sold the world around and its drawings limited only to the size of the crowds that could squeeze into the New Orleans Opera House, on the stage of which they were held.

The Louisiana State Lottery Company was chartered by Louisiana in 1868 and flourished for twenty-five years, in the course of which its contractual arrangement with the state was renewed many times, usually for five-year periods. The golden flood it annually poured into the state treasury paid for many of the improvements that were made in the different parishes. The proprietors of the lottery were widely known, and so, from 1870 to 1888, was the esteemed gentleman who was its manager. Not only did he lend the prestige of his name to the operation, he gave it respectability—which most historians agree he was hired to provide. That he was used as a figurehead or front for the Louisiana Lottery and had little or nothing to do with its operations is painfully evident. He was the ex-Confederate hero, General Pierre Gustave Toutant Beauregard, former superintendent of West Point who during the Civil War had directed the bombardment of Fort Sumter and served with distinction in the battles of Bull Run, Shiloh and Corinth.

On days when drawings were to be made it was part of the ritual for him to take his seat on the stage beside the large revolving glass drum in which thousands of tickets had been dumped and watch the proceedings with an attentive eye. He wore his uniform of spotless Confederate gray on such occasions. As the years advanced he became a pathetic figure as he sat beside the drum watching some unknown Negro lad brought in from the street to pull out the lucky tickets.

Over the years the state's connection with the lottery company had been condemned by certain segments of the population. In 1888, when the company became involved in a scandal growing out of the charge that it had defrauded Louisiana out of several hundred thousand dollars, the lottery became a political issue. The factions opposed to permitting it to continue joined ranks. When General Beauregard suddenly severed his long connection with the Louisiana State Lottery Company, it may or may not have had any relation to the storm that was brewing, but the dissidents hailed it as a victory and redoubled their campaign to keep the state from renewing the lottery franchise for another five years.

No doubt the company spent a great deal of money where it would do the most good and in addition agreed to pay the state $1,250,000 yearly. When the measure came before the legislature it was approved without difficulty. But this was only a brief respite; the federal government had been investigating the lotteries and in 1890 it put them out of business by denying them the privileges of the United States mails. The giant of them all, the Louisiana State Lottery, although badly crippled, managed to continue for three years before it permanently closed. Its demise was followed by the passing of its most famous servant, General Pierre Beauregard.[2]

Exactly one hundred years had intervened between the closing of the Louisiana State Lottery and that day back in 1793 when the commissioners of the District of Columbia endeavored to raise money for the improvement of the city of Washington by means of a lottery. It was not the first lottery conducted in the United States but it was the first and only one sponsored by the federal government. Although undertaken in a good cause, the venture failed to produce the sorely needed funds.

The first canal company to attempt to raise money by a lottery scheme was the Santee and Cooper in South Carolina in 1797. It was not successful and had to be abandoned.

Some years later (1811) the old Patowmack Company, which was building several short canals to improve navigation of the Potomac River around the Great Falls, tried a lottery and was equally unsuccessful. In New England, however, the short South Hadley Falls Canal and the Middlesex fared better. In fact, among the major canals, the Middlesex appears to have been the first to successfully finance itself through its lottery privilege.

Its subsidiary, the mile-long Amoskeag Canal around the falls of the Merrimack River, was the first to become involved in scandals arising from the alleged misappropriation of funds received from the managers of its lottery privilege. Judge Blodgett, the canal's principal promoter, was accused of having diverted to his own pocket $100,000, which was never proven and very likely was not true. Blodgett in turn charged the lottery firm with having cheated the Amoskeag Canal Company out of thousands of dollars. Charges and countercharges filled the air until the matter was finally settled out of court. Blodgett apparently was the winner, for he built a mansion beside the canal and had the satisfaction of seeing it swarm with traffic.

Had the term "racket" been in use with present-day connotations, it would accurately describe the lottery business. It is not surprising therefore that the history of canal companies versus lottery men is one long account of deceit, corruption and larceny, not excluding the Union Canal Lottery of Pennsylvania, the richest of all such concerns. The size of its operations can be judged by the fact that in 1832 alone it awarded $5,216,240 in prizes. Its take for the year was estimated at not less than $20,000,000, of which the canal company received only between 5 and 6 percent, as it reported to the legislature in an appeal for financial aid.

Obviously there was more money to be made in the lottery business than in building a canal. You are entitled to ask why the canal companies did not conduct their own lotteries. There were many reasons, most important of which was that they didn't have a staff of hundreds of trained professional lottery ticket salesmen who could hold their own with the competition. The Union Canal Lottery was reputed to have more than twelve hundred agents spread out over Pennsylvania, Maryland, New Jersey and New York. In a town no larger than Easton, Pennsylvania, there were five ticket brokers competing with each other.

They were hard, glib-tongued men who could recognize an easy mark at a glance. They displayed over the front of their ground floor offices and in newspapers such come-ons as: "Allen's Lucky Boy. Let the Golden Shower Fall on You"; "Greene's Mint. We have the Lucky Numbers"; "Why be poor? Buy Your Tickets at McGowan's." Down at Lancaster "Latshaw's Lucky Office" resorted to verse:

> "If Dame Fortune lets you draw
> You'll find her faithful ever.
> Her only agent is Latshaw,
> And he'll forget you never."

In Philadelphia the S. & M. Allen Company said simply but persuasively, "We have the right numbers. Six big winners the past year."

By the end of 1831 independent brokers dealing in the tickets of rival lotteries had practically been put out of business by the Union Canal Lottery Company. It had come into existence back in 1811, when the Schuylkill and Susquehanna Canal and the Delaware and Schuylkill Canal, with legislative approval, were merged into the Union Canal and granted a renewal of their original lottery privileges. As discussed in an earlier chapter, the state fondly hoped that the new arrangement would provide sufficient funds for completion of the joint waterway. It provided funds in abundance, but for the lottery company, not for the Union Canal, which was completed only by extensive grants from the state.

Over the years between 1811 and 1833 the Union Canal Lottery Company conducted almost fifty lotteries (an average of two a year) and distributed $33,000,000 in prize money—a great share of which it paid to itself. This was accomplished by withholding thousands of tickets from sale and when a drawing was about to occur, dumping the stubs into the drum along with the stubs of the tickets the public had purchased. Naturally the company stubs won many prizes. Eventually this utterly fraudulent trickery had a great deal to do with putting the Union Canal Lottery Company out of business.

Wealthy men seldom invested in lottery tickets; it was the small farmer and the laboring man, dreaming of winning first prize ($20,000), who made the enterprise fantastically profitable for the proprietors. If a man bought only a ticket or two, he couldn't be hurt much; but there were fools who, determined to force their luck, wagered their entire savings. When they lost, the results were sometimes tragic.

The total prize money amounted to $80,000, which was divided two-hundred ways, scaling down from the ten grand prizes to several hundred prizes of $5 and $10. The price of a full ticket in a Union Canal Company Lottery was $5, but to enable the very poor to participate, tickets were divided into eighths, which meant that for as little as sixty-three cents you could participate in a drawing. The odds against winning were astronomical. But some people did win, of course, and the company was delighted to publish their names and give them wide publicity.

The drawings of the Union Canal Lottery were held on the landing at the head of the main stairway leading to the second floor of the State House at Harrisburg, which gave the proceedings an aura of governmental approval. They were attended by masses of people, jammed together on the main floor, groaning, cursing or cheering as their numbers proved lucky or unlucky.

The January drawing in 1833 turned into a riot when a loser drew a gun and killed a bystander before committing suicide. A woman who had lost her all became a raving maniac. Newspapers throughout the state joined ranks in demanding that the lottery—all lotteries—be abolished. The crowning argument seems to have come when the Union Canal reported that it had received only $60,000 of the $1,000,000 or more the lottery company had taken in.

After refusing to face up to the question for years, the legislature now acted promptly and made conducting a lottery a crime under the laws of the state of Pennsylvania. Banned now in Pennsylvania as well as New York, the promoters of such enterprises turned their attention to unloading the stock of paper railroads on the public, which for several decades proved to be a profitable substitute for lottery profiteering.

25

Fast Packet For—

IN THE DAYS when life was leisurely, traveling by canal packet must have been a pleasurable experience. If weather was favorable you could sit on the cabin roof with the country spread out in all directions around you. If you did not reach your destination before nightfall, after supper the main cabin was transformed into a crowded and uncomfortable bedroom, with a curtain put up to provide a measure of privacy for the female passengers.

On the better packets food was uniformly plentiful. Caustic Charles Dickens, the famous English novelist who saw little to his liking on his visit to America in 1842, found the meals on the Pennsylvania Main Line Canal more than adequate. "The menu for breakfast and supper," he wrote, "offers the traveller tea, coffee, bread, butter, salmon, shad, liver, steaks, potatoes, pickles, ham, chops, black-pudding and sausages. Midday dinner was precisely the same, only minus tea and coffee." How such a variety of food could have been prepared for from thirty to forty passengers in a space as tiny as a packet-boat galley passes comprehension.

Dickens's description of the architecture of a packet is accurate and may have been copied from a popular *Travelers' Guide* of the day.

It resembles a small Noah's Ark—a houseboat whose only deck is the roof of the cabin. In the bow, carefully cut off from the rest of the boat, is a tiny cuddy for the crew. Next back of this is the ladies' dressing room and cabin, sometimes a separate room, sometimes cut off from the main cabin only by a red curtain. Next is the main cabin, 36 to 45 feet long, which was saloon and dining room by day and men's dormitory by night. Back of this is the bar, and finally, at the very stern, is the kitchen, almost always presided over by a negro cook, who is usually the bartender also. The other members of the crew are the captain, two drivers, two steersmen, one each for the night and day shifts.

It was a time when prominent figures of the literary world, male and female, American as well as British, were engaged in profitable lecture tours. Of necessity, much of their traveling had to be done by packet. Without exception they earned money between engagements by writing articles for such magazines as the *North American Review,* describing the changing scene they encountered from day to day and the delays and exasperations to be met with on the canals.

Of the touring American celebrities, New Englanders to a man, only William Cullen Bryant found canal travel enjoyable; to Whittier it was "a trying experience"; Hawthorne found a daylight journey by canal preferable to being bounced around over unimproved roads in a crowded stage. "But by night," he observed, "the main cabin becomes a chamber of horrors in which sleep is impossible."

Dickens was equally aghast at the indignities inflicted on the passengers every night at ten o'clock, when they were herded up on deck while the main cabin was being transformed into a communal bedroom. The nightly performance began, he wrote:

. . . when two or three members of the crew began carrying adjustable berths, sheets, pillows, curtains, and so forth, into the main cabin. . . . Each berth was a narrow wooden or metal frame with a strip of canvas fastened over it. It was held in position at one side by two projecting iron rods which fitted into two holes in the side of the cabin; and on the other or front side by two ropes attached to the edge of the frame and suspended from hooks in the ceiling. There were at least three beds in a tier, one above another, sometimes four; and all fastened to the same rope. The tiers were set as closely together as possible all around the cabin, which thus furnished beds for from thirty-six to forty-two people. It seems incredible that so many people could have lain down in the limited space on one of those boats, even with the floor and tables covered with them.

Dickens goes on to say that when he first went below at ten o'clock and caught his first glimpse of the tiers of hanging bunks he mistook them for:

hanging book shelves designed apparently for volumes of the small octavo size. Looking with greater attention at these contrivances (wondering to find such literary preparation in such a place) I decried on each shelf a sort of microscopic sheet and blanket; then I began dimly to comprehend that the passengers were the library, and that they were to be arranged edgewise on these shelves till morning.

The method of assigning berths varied on the different canals. On the Pennsylvania Main Line, on which Dickens was traveling at the time, the choice was by lot.

> I saw some of the passengers gathered around the master at one of the tables, drawing lots with all the anxieties and passions of gamesters depicted in their countenances; while others with small pieces of cardboard in their hands were groping among the shelves in search of numbers corresponding with those they had drawn. As soon as any gentleman found his number he took possession of it immediately by undressing and crawling into bed. The rapidity with which an agitated gambler subsided into a snoring slumberer was one of the most singular effects I have ever witnessed. As to the ladies, they were already abed behind the red curtain, which was carefully drawn and pinned up the center; though as every cough or sneeze or whisper behind the curtain was perfectly audible before it, we had still a lively consciousness of their presence.

Frances Trollope, Dickens' countrywoman, the mother of the well-known novelist Anthony Trollope and a writer of some consequence herself, returned to America after the death of her husband, Thomas Anthony Trollope. She was in the United States gathering material for her book, to be entitled *Domestic Manners of the Americans.* She had little reason to regard the country with affection, her husband having capped a succession of business misadventures with the failure of what was called a "fancy goods shop" in Cincinnati.

Mrs. Trollope had occasion to travel by canal from Schenectady to Buffalo. As described in her book the voyage was a horrendous experience, relieved only by the short side trip from Utica to Trenton Falls, which she found "beautiful and inspiring."

Of her experience on the Erie, she remarks, "I can hardly imagine any motive of convenience powerful enough to induce me again to imprison myself in a canal boat under ordinary circumstances."

The sweltering heat of the cabin, the hordes of mosquitoes and the inaccuracy of the spittoon-users, which so offended Mr. Dickens, drew her fire but it was the women aboard who irked her most.

"The accommodations being greatly restricted," she complains:

> everybody, from the moment of entering the boat, acts upon a system of unshrinking egotism. The library of a dozen books, the backgammon board, the tiny berths, the shady side of the cabin, are all jostled for in a manner to make one greatly envy the power of the snail; at the moment I would willingly have given up some of my human dignity for the privilege of creeping into a shell of my own. To any one who has been accustomed to traveling, to be addressed with "Do sit here, you will find it more comfortable," the "You must go there, I made for this place first," sounds very unmusical.

She remarks on the beauty of the Oneida and Genesee country, but she concludes, "Had we not returned by another route [by stagecoach] we should have known little about it. From the canal nothing is seen to advantage, and very little is seen at all."

Philip Hone, a polished gentleman, several times mayor of New York City and president of the Delaware and Hudson Canal Company, had an entirely different view of his journey on the Erie in 1835, long after Mrs. Trollope rode Erie water. He wrote in his diary:

> The boat was not crowded, the weather was cool and pleasant, the accommodations good, the captain polite, our fellow-passengers well-behaved, and altogether, I do not remember to have ever had so pleasant a ride on the Canal. My hammock, to be sure, was rather narrow and not very soft and my neighbor overhead was packed close upon my stomach, but I slept sound as a ploughman and did not wake until tapped on my shoulder by the boy and told to "clear out."

Twelve years later, however, found him more or less agreeing with the sharp-tongued Fanny.

"This canal traveling is pleasant enough by day time, but the sleeping is awful. . . . The sleepers are packed away on narrow shelves fastened to the side of the boat, like dead pigs in a Cincinnati pork warehouse. We go to bed at 9 o'clock, and get up when we are told."

When Mrs. Trollope's book appeared, it aroused the ire of Americans. But they bought it. In fact, Fanny Trollope set a pattern that was followed by Helen Martineau and other English writers which more or less holds good today: Hold us up to scorn for our bad manners, or lack of manners, and we'll beat a path to the bookstore.

Some of our own lady novelists tried to fight back, notably Caroline Gilman, a prolific producer of peach-and-magnolia fiction. She went into ecstasies about the Erie and the Champlain canals, breaking into poetry from time to time. Hawthorne dismissed her as "one of that damned mob of scribbling women." The fact that she was a pro-slavery southerner and he a New England abolitionist may have had something to do with his cavalier treatment of her.

During the summer she spent on the canals of New York State she kept a journal, much of which was published later. "Are the scenes I have witnessed," she asks, "really among the same population which English travelers have described? Am I dreaming, when I find only courtesy among the cultivated and quietness among the other classes?" Perhaps Miss Gilman deserved better from Mr. Hawthorne than she received. But of course he and the other Boston Brahmins were more than a little pro-British at heart.

The days of the Old Erie ended when the last stretch of "the new and improved Erie" from the Hudson to Rome, with its thirteen-foot deep channel, larger locks and vertical drop gates, was completed. It marked the end of the era with which most of the balladry and literature of the waterway is concerned. By 1855 canals everywhere were fighting a losing battle with the railroads; for every mile of channel that was being abandoned, ten miles of track were laid. The days when touring politicians addressed canalside gatherings from the roof of a packet were passing; they were now dashing about the country making whistle-stop appearances from the rear end of a passenger train.

Out in Ohio and Indiana the arrival of a packet, bringing the mail and news of the outer world, was still the big event of the day. Local boys were always on

On the "horse" boats, the stable was always in the bow. *Courtesy New York Public Library Picture Collection.*

hand, gazing enviously at the young hoggee who, aware of his audience, disdainfully exhibited his skill in the use of the long whip.[2]

In canal parlance horse or mule could refer to one or the other, the term being used without distinction. But mixed teams were regarded as unsatisfactory and were seldom used. On the Ohio and Erie, although it was a long canal, there were very few relay barns stocked with fresh animals, it being the general custom for a boat to carry a relief team aboard. When Captain Parkhurst of Circleville, whose boat *The Ark* was widely known as "the Sunday boat" because it did not

A popular attraction on the Wabash and Erie Canal for many years.

run on Sunday, was asked why he carried a spare team, he replied, "It saves time and it's cheaper."

Asked if it wasn't hard on the horses, he shook his head.

"It's easier, I'd say. I keep a team on the path for four hours, then bring 'em aboard and put the other team out. That gives time for a good rest. On this boat they have all day Sunday off. I turn 'em out on grass if there's any to be had. If a man uses any judgment a good mule will last as long as a good boat."

"How long would that be, Captain?"

"Eighteen to twenty years—depending some on the weather. A stiff head wind blowing all day holds a boat back and makes it harder on the mules."

This evaluation seems to have been the consensus among canal boat captains. It does not erase the fact that many animals were overworked and abused. On the other hand, the story is told of a mule that had seen long service on the James River and Kanawha Canal. After being retired and turned out to pasture for a few weeks the mule was following the hounds, leaping over fences and leaving blooded horses taking part in the hunt far behind.

For years the Miami and Erie Canal was the favorite hunting ground of traveling medicine shows. The boat was so built that one side of the cabin could be lowered to a horizontal position to make a platform on which the performance took place. The company invariably consisted of a pair of agile young Negroes who could sing, dance and hit fancy notes on a banjo, a tenor to jerk a tear with a soggy ballad and a blonde soubrette, usually the "doctor's" wife or mistress. And always a shill or two to mingle with the crowd and start the parade to the platform to purchase whatever nostrum was being hawked. Often it was a liniment "good for man or beast," or a cure for the ague which was heavily spiked with alcohol.

Then there was the "painless" dentist, whose visits were often eagerly awaited. In the wavering light of the gasoline flares while banjos twanged a lively tune, the extractions took place, the charge a mere fifty cents.

On the Wabash and Erie, the Spalding and Rodgers Floating Palace Circus operated for years, presenting a moth-eaten lion and assorted small animals, several clowns to amuse the children, trained ponies and a death-defying high-wire act spanning the canal.

The most important canalside circus to appear on the Erie was the Sig Sawtelle Show. A flotilla of three boats was required to move it. No performances were given during the winter months, but after tying up at one or another of the larger towns such as Rochester or Syracuse, its menagerie was kept open and attracted some business.

By the mid-1830s many of the canal towns had (using the term loosely) what could be called a theatre. The theatrical fare they offered was largely confined to the presentations of groups of amateur thespians. Several of these groups toured the Erie during the summer months until the invasion of touring companies of professional actors, starring the leading personalities of the American and English stage, including the elder Junius Brutus Booth and the great Irish actor Tyrone Power, made it unprofitable to continue.

Unquestionably canal life was changing, more noticeably on the Erie than elsewhere. There was more religion, more churches, more Irish and more marriages as hundreds of young colleens crossed the Atlantic to wed the sons of the original bogtrotters. An athletic competition or brawl was still regarded as the highest form of entertainment. But the old stories and old songs were being forgotten.

That the Chesapeake and Ohio, the Pennsylvania Main Line and the canals of Ohio and Indiana had their own folklore and ballads must be taken for granted. But they have not been preserved. However, the folklore of the Erie has been captured in type along with much of its balladry, the latter partly because of the pioneer work of John and Alan Lomax. Equally important has been the research and discoveries of Professor Lionel D. Wyld.[3]

The present vogue of country and western music makes one wonder how long it will be before some of the leading performers of that type of song discover canal music. Its themes are seldom synthetic and more often than not indigenous. Among the most popular were: "The E-R-I-E" (accent on final E), "The Raging Canal," "Low Bridge," "Everybody Down," "Ballad of John Mueller and the Lock Tender's Daughter" and "The Girl from Yewdall's Mill." The lyrics are banal and often crude, but they were written to be sung, not read as poetry.

On a quiet summer's night, with twenty to thirty boats lined up awaiting their turn to get through a lock, it must have been something to hear when a bearded bowman began singing "The Raging Canal" (of which there were fifteen verses) at the top of his lungs with the crew of every boat in the line joining in.

The lyrics were lugubrious, which was the case with most canal songs. Undoubtedly they were written by many different men and added to or changed from time to time. The Irish were responsible for many of the songs, so it was not strange that they were sung to old Irish tunes.

Here are a few verses of "The Raging Canal":

> *Come listen to my story, ye landsmen, one and all,*
> *And I'll sing to you the dangers of that raging Canal;*
> *For I am one of many who expects a watery grave,*
> *For I've been at the mercies of the winds and the waves.*

I left Albany harbor about the break of day,
If rightly I remember 'twas the second day of May:
We trusted to our driver, altho' he was but small,
Yet he knew all the windings of that raging Canal.

It seemed as if the Devil had work in hand that night,
For our oil it was all gone, and our lamps they gave no light,
The clouds began to gather, and the rain began to fall,
And I wished myself off of that raging Canal.

The Captain told the driver to hurry with all speed—
And his orders were obeyed, for he soon cracked up his lead;
With the fastest kind of towing we allowed by twelve o'clock,
We should be in old Schenectady right bang against the dock.

But sad was the fate of our poor devoted bark,
For the rain kept a pouring faster, and the night it grew more dark;
The horses gave a stumble, and the driver gave a squall,
And they tumbled head and heels into the raging Canal.

The Captain came on deck, with a voice so clear and sound,
Crying cut the horses loose, my boys, or I swear we'll all be drowned;
The driver paddled to the shore, altho' he was but small,
While the horses sunk to rise no more in that raging Canal.

No men employed on the Erie had a wider acquaintance than the canal walker, who covered ten miles a day trudging along the towpath—five miles to east or west and five miles back to the point of starting. It was his job to watch for leaks that developed in the locks, examine the waste weirs and report signs of an impending break in the channel bank. He was welcome at the saloons along the way for he brought the latest news and scandals of the day, and in the process picked up tidbits to add to what he already knew.

Canal walkers had time aplenty to invent tall tales of their own and fit new lyrics to old tunes as they walked off the miles. It was a canal walker who invented the story of the captain who bought a dead whale that had been washed up on a Cape Cod beach and set out to get rich by exhibiting it. After being exposed to the sun for ten days the decaying whale became so odoriferous that no one would come within a hundred yards of the boat. To get rid of the decaying carcass the captain tried to dump it in the canal, only to be caught and fined $2 by a justice of the peace for interfering with traffic. Desperate, he finally succeeded in convincing a farmer that the whale would make excellent fertilizer and knocked it down to him for eight cents.

26

Canal Sin,
Violence and Prosperity

T HE LABOR OF five thousand men was required to build the Erie Canal. Ten years after coming into operation, its facilities taxed to capacity by the vast amount of business it was handling, the canal was employing no fewer than twenty-five thousand men (and a few women). That figure does not include the thousands, not employees, whose livelihood was derived from the prosperity of the Erie. No other canal or canal system ever approached the Erie in the amount of tonnage forwarded or the size of its payroll. Combine the best yearly figures compiled by the Chesapeake and Ohio, the Pennsylvania state canals and the Ohio and Erie and they fall far short of equaling the best years enjoyed by the Old Erie.

"As a canal it is in a class by itself," wrote Captain Frederick Marryat, the English journalist. "Travel on the Erie is made lively by the strife and bickering between boaters and canal employees which occurs at every lock."

This and similar observations undoubtedly indicate why so much of our canal folklore has an Erie background. Other canals had their canallers; the Erie had its *canawlers*—a semi-vagabond breed of hard-living, hard-fighting, hard-drinking men, who regarded all rustics and land-bound men with contempt. They were numerous enough to live their lives among their own kind, suspicious of outsiders, asking no favors and giving none. Their women, whether born and raised on the canal or runaways from the drudgery and monotony of the farm, shared their sentiments and animosities.

The great majority of the canawlers were not native New Yorkers; the prospect of good wages and a free and easy life on the canal had lured them into New York from other states along the Atlantic seaboard. They found the region through which the Erie coursed westward still thinly settled, only one generation removed from being frontier country at the mercy of Iroquois outrage.

The pioneers who had subdued the Indians and carved their farms out of the unmapped, timbered wilderness they called the North Country were gone.[1] But their prosperous sons and grandsons were to be found in every Erie Canal town and village. They had helped to build the Erie, and possessing the ambition and perspicacity of their forebears, had easily made the transition from tillers of the soil to the world of business. As shippers, merchants, line-boat agents and owners of grain elevators they were in constant contact with canal captains and canawlers. The latter, with their suspicions and hostility, looking for injustice and trickery where they did not exist, made it an abrasive relationship. Over the years they were responsible for much of the violence that occurred on the Erie. When the canawlers wearied of fighting the so-called monied interests, they fought each other.

Fishing the floating body of a dead man out of a canal was not an unusual experience. No one has attempted to compile a record of how often it happened, but on the Erie Canal more than two score floating bodies were recovered. Often a knife or gunshot wound was evidence of how the deceased had come to his death. In other cases the coroner's verdict was "death by drowning," and seldom was any attempt made to discover whether the drowning was indeed accidental or whether the deceased had been knocked on the head, robbed and then rolled into the canal.

It is significant that 75 percent of these unexplained drownings or killings associated with the Old Erie occurred in the vicinity of Watervliet, where the infamous Side-Cut connected the canal with the Hudson and the city of Troy. The Side-Cut, dubbed by writers "the Barbary Coast of the East" in the 1880s, with its reputation for "a hundred fights a day, a body a week in the canal," was the recognized citadel of crime, violence and debauchery. In the space of two city blocks it flaunted twenty-nine saloons, with two exceptions low groggeries with retinues of overage whores imported from New York City in attendance. Among these women, the discarded refuse of a great city, were criminals of every degree of depravity, ready at a moment's notice to carry out the commands of their pimp masters.

For some inexplicable reason canal historians and novelists have refused to recognize that there was sex on the Erie Canal. Samuel Hopkins Adams acknowledges its presence in several ribald anecdotes but goes no further; Walter D. Edmonds, a masterful writer who has given millions of Americans their only glimpse of canal life, sidesteps the matter gracefully in *Rome Haul*. In the scene in which Molly Larkin, his heroine, comes aboard Dan Harrow's boat for the first time, ostensibly to be his cook, he leaves no doubt that she is to be his mistress as well. For two hundred pages she remains aboard the *Sarsey Sal*, sleeping in the tiny cubby with Dan, sharing his bed or sleeping in close proximity. But Edmonds meticulously avoids revealing what their extramarital relations were. Salty Sam Adams, a true Erie product, born and raised on the canal, was equally careful to leave the sex life of his characters to the imagination of his readers.

Of course during the first third of this century writers labored against strictures that have since disappeared. Undoubtedly that explains why in canal literature, especially as it concerns the Erie, so little is said about the cruelty, depravity and violence that were a part of canal life. That the American Tract Society, the Seaman's Friend and Bethel Society regarded the Erie Canal as the very hotbed of sin and iniquity is attested by the hundreds of missionaries they dispatched to the Erie and kept there year in and year out.

Most of the missionaries were young men. Their sincerity was not to be doubted, but being professional agents of the Lord, they wrapped themselves in a robe of sanctimoniousness that awakened the contempt of men whose souls they were there to save. But they persevered and rescued many men from disgrace and ruin. Demon Rum was the principal enemy. Whisky was so cheap—a "fip" (6¼ cents) the half pint—and blackstrap (rum and molasses) was 3 cents the glass, so it was small wonder that boaters drank to excess.

Harlow quotes an old missionary as saying, "It is a sad sight to see the evils of rum among the boatmen. . . . They not only drink it at the rum shop, but carry it aboard their boats." The rum sellers were willing to admit that the sale of rum produced vice, poverty and wretchedness—but they were unwilling to discontinue the vile traffic while it continued to be profitable.

The like of the notorious Side-Cut was not to be found elsewhere, but the red-light districts of Albany and Buffalo were hardly less notorious for their crime and depravity. There was a dram shop adjacent to most locks, often owned by the lock-tender. At the relay stations (the horse barns) a groggery was always nearby, its nightly hilarity enlivened by several overage resident prostitutes. In such places, according to the sentimental nonsense of canal balladry, "the jolly boaters danced and caroused the night away."

To paraphrase Gilbert and Sullivan, the lot of the canal missionary was not a happy one. Unquestionably they won their greatest victory when they so aroused public opinion that the captains and boat owners had to treat their young drivers, hoggees, boys whose ages ranged from twelve to seventeen, more humanely. By 1850 more than a thousand youths were walking or riding the towpath.

For no better reason than that all were poor, they are usually lumped together as an Erie product, one indistinguishable from the other. Actually they sprang from many different backgrounds—farms, city slums, country villages, and no few were the sons of the Irish bogtrotters who had dug the Erie. After a year on the canal most of them were hard and tough, masters of profanity, and, where custom permitted, could stand up to a bar and down their whisky as neat as any man.

They were on the towpath every day, no matter what the weather. They had to be expert in dropping or raising towlines to permit the passage of fast packets and freighters. If there was a foul-up, the captain vented his wrath on them, which often meant a whipping and/or the withholding of their pay. The standard wage for hoggees ranged from $8 to $10 a month for a season of seven months, payable when the boat tied up for the winter. It was an arrangement that worked in the captain's favor, for a boy could not jump ship without risking loss of the wages he had coming.

The missionaries denounced the custom of paying off boys at the end of the season as a vicious practice, claiming (and rightly) that putting $60 to $70 in the hands of a youngster craving excitement made him an easy target for harpies and the canalside underworld. They cited numerous incidents of the hoggees being robbed when they entered such hell-holes of vice as the Side-Cut.

In canal fiction the hoggee occupies a prominent position. Especially is that true in the juvenile field. Writers of the Horatio Alger rags-to-riches school found in the poor, mistreated hoggee, striving to make his own way, an ideal figure on which to base their tales. Which is not to suggest that ambitious, thrifty boys of character, determined to get ahead, were not to be found on the canals. Every canal port and town could point with pride to one or two of its prosperous elderly citizens who had begun working on the Erie at thirty cents a day. Michael Moran of Frankfort was an outstanding example. He was only fifteen when, newly arrived from Ireland, he began as a hoggee on the Erie in 1850. Three years later he had advanced himself to steersman, the goal of all drivers. By 1857 he had saved enough money to buy his own boat, the *Cayuga*, and became his own captain.

The *Cayuga* was a small boat and an old one, but she was only the first of many boats he was to own. To celebrate his good fortune, he took time off to take out his citizenship papers, something he had been too busy to do in the past.

He must have paid strict attention to business and given shippers a superior brand of service, for in 1860 there was a small fleet of Moran-owned boats making the Erie run from Albany to Buffalo. It was only the beginning of the long trail Mike Moran, his sons and grandsons were to travel. Today, if you have occasion to cross New York harbor, you will be struck by the number of squat, grunting tugs, many of them ocean-going, that you will see escorting Atlantic liners to their berths or ferrying fleets of scows bearing whole trains of freight cars to their various destinations—each decorated with a big block M on their stack, the house emblem of the Moran Towing Company.

Another Irish lad destined for fame, Paddy Ryan of county Tipperary, arrived in Troy in 1859 from the auld sod, accompanied by his parents. He was eight years old at the time. At eighteen he was employed as a lock-tender. Lock-tenders were anathema to most captains and their brawling crews. To survive, he had to depend on the prowess of his fists. Apparently he had little difficulty disposing of the bullies who tried to impose on him. His spreading reputation as a fighter brought him to the attention of Jimmy Killoran, the Troy boxing master and promoter. Killoran taught him the so-called manly art of self-defense and made himself young Ryan's manager.

Since Ryan was an authentic canal hero and closely associated with the Erie in the public mind, his ring career warrants a brief inclusion in this narrative. Killoran brought Ryan along carefully, matching him against increasingly important fighters. Paddy Ryan ran up a string of victories, and the purses became larger and larger. It was a time when ownership of a super-elegant saloon was deemed unfailing evidence of a prizefighter's greatness and affluence. Thus Ryan opened a saloon on the Side-Cut where his presence behind the bar was the chief attraction. The establishment was immensely profitable, for he was, as the *Sentinel* had proclaimed, "The Pride of Troy."

For a time it seemed there was no limit to his good fortune. Danny Dwyer, the reigning American champion, had been matched to fight Joe Goss, the English champion. When illness forced Dwyer to retire from the ring, Paddy Ryan was substituted for him. On June 1, 1880, he faced the English champion in a scheduled bare-knuckles fight to a decision, fought under London Prize Ring rules, with a knockdown being scored as a round.[2]

Many states had banned prizefighting. To outwit the authorities, fight promoters kept the actual scene of battle a secret until the last minute. When Ryan stepped into the ring only several hundred spectators were in attendance. It took him eighty-five rounds to put the aging Goss away. A dispatch in *The New York Times* the following day gave the details:

"A prize fight for the heavyweight championship and a side bet of $2,000 took place near Colliers, West Virginia, in a ravine 300 yards from the Pennsylvania boundary line. It went 85 rounds, with Paddy Ryan being declared the winner. Actual time of the fight was one hour and 27 minutes. Only about 300 spectators were in attendance, due to poor transportation."

His reign as champion was brief. John L. Sullivan, the Boston Strongboy, was soon on his trail. Ryan fought Sullivan three times and was always the loser, making such a poor showing that the late Quentin Reynolds was led to write, "Paddy Ryan spent most of his time in the ring with John L. Sullivan—usually on his back."

If Watervliet had the violence and hurrah of the Side-Cut to contend with in one direction, it had the dignity and peacefulness of a large Shaker community from the other. The advance guard of the disciples of "Mother" Ann Lee, the founder of the cult, had settled in New York State the year after the completion of the Erie Canal. They were an offshoot of the prosperous Shaker colony Mother Lee herself had founded at Enfield, Connecticut, shortly after quitting England for the United States. Only a handful of seven or eight adults when they settled at Watervliet, by 1850 they numbered well over a thousand and had established a string of Shaker settlements. Their adult converts, like themselves, had taken vows of celibacy; the children among them had, for the most part, been rescued from the horrors of orphan asylums.

The soil of New York State seemed to provide a special kind of nutriment that enabled any cult, religion or "community" to prosper. Some of them are still with us, such as the Oneida Community and Church of Jesus Christ of Latter-Day Saints (the Mormon Church). You will recall that it was on the Hill of Cumorah near Manchester in Ontario County, New York, that Prophet Joseph Smith is said to have found the gold plates of the Book of Mormon.

Almost from the time that a canal began operating, churchmen of various denominations joined ranks in demanding that travel be banned on the Sabbath. The protesters cited instances of a score or more boats blowing their horns and bugles as they jockeyed for position at a lock, disrupting holy service in nearby churches. In Massachusetts the Middlesex Canal responded by prohibiting all Sunday travel between Boston and Lowell. In Pennsylvania the state-owned canals swung into line by permitting only through packets to proceed on the Sabbath. Great pressure was exerted to limit Sunday traffic on the Erie. Many owners and captains voluntarily tied up their boats on Saturday evening. Rochester passed an ordinance prohibiting the blowing of canal horns within the town limits on Sunday. From first to last the privately owned Delaware and Hudson was a strict Sabbath-keeping canal. Its locks closed at Saturday midnight and did not reopen until Monday morning. This often resulted in twenty-five or more boats being gathered at a lock. Missionaries were always on hand to conduct impromptu services.

As a class boaters were overly profane. There were exceptions of course. A notable one was Captain Abram Spurling, known up and down the Erie as "Holy Moses" Spurling, for no matter how great the provocation, his profanity was limited to that expletive.

What used to be known as Erie Country produced a surprisingly large number of young men who were to leave their mark on America. Many of them were connected in one way or another with the Erie Canal before they packed their bag and by packet boat journeyed to Buffalo and on westward to become, in the vernacular of the times, "Princes of Industry." Commodore Vanderbilt cannot be included in this category, for he was an outsider, although he did much to shape the history of Central New York with his railroads. Among the others were John Davison Rockefeller who left his home town of Richford, in Tioga County, and journeyed by the Erie Canal to Buffalo and on to Cleveland, where on $900 of capital borrowed from his father he launched himself in the commission business, handling farm products. (The oil refineries came later.) George Mortimer Pullman set out from his home village of Brocton, New York, to become the sleeping-car king.[3] The aforementioned William Butler Ogden, a young country lawyer, set out from Walton, New York, for booming Chicago, which is his debtor even today. And there was Frank Parmalee. Any reader who ever changed trains and depots in Chicago, will remember the Parmalee buses—"Our agents meet all trains."

Parmalee, the son of a struggling farmer, was born on the Erie Canal at Port Byron, a village that produced more than its share of noted men. After working on the canal for a year, he sailed the Great Lakes as purser on various steamers, saving his money and looking for an opportunity to get ahead. He found it in Chicago in 1853, when he launched the Parmalee Omnibus and Transfer Company. Three years later he gave Chicago its first street railway.

When you examine the beginning and development of that great multi-million-dollar convenience known as the express business, which in some way touches the lives of all of us, you discover that five of the six pioneers who established it were related to the Erie Canal.

William F. Harnden is usually credited with having originated what came to be known as "the express business." Actually he was preceded by the brothers L. B. and B. D. Earle, all three Massachusetts Yankees. They were private messengers competing with the slow and wretched United States mails of the time (the mid-1830s) rather than expressmen. They literally carried their business under their tall beaver hats, for it was limited to the safe and speedy transmission of business communications between the bankers and merchants of Boston and New York, which was accomplished in eighteen hours by way of the pioneer Boston and Providence Railroad to Providence and Long Island Sound steamer to New York City, roughly a third of the time needed by the post office.

Harnden's business soon outgrew the beaver hat. He took in his brother Adolphus as a partner and they began carrying voluminous carpetbags. But that didn't suffice for long. They next graduated to the steamer trunk, which they somehow always managed to keep in sight. In 1839 William Harnden began operating a triangle route from New York to Albany to Boston. The following year his borther Adolphus perished, along with $40,000 entrusted to his care and all his fellow passengers, in the tragic burning of the steamship *Lexington* off the Long Island shore.

In 1841 Harnden engaged a young man by the name of Henry Wells to be his agent in Albany. Young Wells had a stammer and to partially conceal the

Henry Wells, the great expressman, a product of "Erie Country." *Courtesy Wells-Fargo Bank, History Room, San Francisco.*

William Fargo, Wells's equally famous neighbor and partner. *Courtesy Wells-Fargo Bank, History Room, San Francisco.*

twitching of his facial muscles when speaking he wore a luxuriant brown beard. He had been employed on the Erie for several years as a forwarder of freight and passengers at Syracuse. His head was filled with grandiose ideas of what the express business could be. He urged extending service to Buffalo, Chicago, Cincinnati and St. Louis. As he saw it, any place that could be reached by canal, railroad, steamboat or pack train was potential express company territory.

Harnden, now sole owner of what had become a very profitable business, was not impressed by such wild talk and he soon dismissed Wells in favor of a more conservative man. It was the greatest mistake of his life, for Henry Wells was to go on to fame and fortune as one of the giants of the multimillion-dollar express business.

Wells, a Vermonter by birth, the son of a Congregationalist minister, grew up in what with the coming of the Erie Canal became Port Byron, a few miles west of Syracuse. For a time he was the village cobbler, but when the canal, far from being completed, was opened to navigation between Schenectady and Rochester and the boats began discharging freight consigned to merchants in neighboring towns, it occurred to Wells, as it had to several other alert men, that transporting such shipments, and passengers as well, from canalside to their destination could be developed into a profitable business. His friend George Pomeroy was operating a stage line north from Syracuse to Oswego; James Wasson was running a tri-weekly stage to Auburn; south from Syracuse William Fargo, in connection with his livery business, was competing with John Butterfield for the stage business. These five men—Henry Wells, William Fargo, John Butterfield, George Pomeroy and James Wasson—living within an area of a hundred square miles, were to be the founders and guiding heads of the great coast-to-coast express business of the United States. Crawford Livingston, scion of the influential New York City family of that name, joined them in the days when they were organizing and conducting their business under a succession of different corporate names. Only Alvin Adams, the founder of the Adams Express, was never connected with the Erie Canal.

It seems to have been agreed long ago that no chronicle dealing with Erie Country and its expressmen could be considered complete unless it dwelt at some length on the arrival at Laidley's Restaurant in Buffalo of the first shipment of fresh oysters ever to reach that city. It is not surprising that in its many tellings the facts have become garbled. Some say that the idea originated with Commodore Vanderbilt and that he used it to promote the sale of stock in his New York Central Railroad. Lucius Beebe and Charles Clegg, Wells Fargo's most knowledgeable historians, say otherwise.[4] According to them the bivalves were rushed up the Hudson to Albany on one of Dan'l Drew's steamboats, where they were shucked, placed in containers and sped westward by train to Batavia, the end of the track, from where Pomeroy and Company (Henry Wells was the company) got them to Buffalo in jig time.

Beebe and Clegg give full credit to Henry Wells for conceiving the idea and carrying it out successfully. The passing years were to disclose that he had a large repertoire of original ideas.

There were other New Yorkers, not connected with the express business and only faintly related to the Erie, who were to change the life-style of many Americans. Such a one was Wells' neighbor in Port Byron—Isaac Singer. No one seemed to know quite what to make of him; his head was in the clouds and he seemed to spend his life just tinkering. But in 1851 he perfected and patented what he called his "sewing machine" and made women the world over his debtors.

The Erie Canal of a bygone era has lost its identity; but the blessings it bestowed on the country through which it flowed are everywhere discernible. And the names of the men who contributed to its greatness or were made great by it stand out indelibly in Erie lore.

Notes

CHAPTER 1

1. New England's most popular and widely grown apple, the Baldwin, was named for Loammi Baldwin.

2. Many writers have taken the liberty of dropping the *k* in spelling Merrimack. If you consult your atlas you will find that the town and county of that name in New Hampshire, as well as the river, are spelled Merrimack.

3. Loammi Baldwin, *Letters* (Mass.: Medford, Woburn, Worcester Historical Societies, 1800–1803).

4. Christopher Roberts, *The Middlesex Canal.*

5. John Langford Sullivan bought a marine engine from Oliver Evans, the inventor of the high-pressure engine in America—an honor erroneously bestowed on Robert Fulton—installed it in a small vessel designed for towing purposes and experimented with it on the Middlesex Canal. Fulton came to see it and, discovering that Sullivan had not applied for a patent, did so at once. This led to a series of lawsuits. The issue was decided in Sullivan's favor, but he was never able to compete with the powerful New York monopolists backing Robert Fulton.

6. When the Boston and Lowell Railroad was chartered, it sealed the fate of the Middlesex. Caleb Eddy, its last superintendent and a capable man, realized it. "The future has but a gloomy prospect," he wrote in 1843. He conceived the idea of making the canal into an aqueduct to supply Boston with fresh water, for although the city now had a population of 100,000, its inhabitants were still getting their drinking water from wells and cisterns. Supported by a number of leading doctors and chemists, Eddy argued that impure water was responsible for much of the sickness in Boston. Undoubtedly this was true, but the city fathers rejected his proposal out of hand.

CHAPTER 2

1. The worst accident in the history of the Connecticut River occurred at 3:00 A.M. on the morning of October 9, 1833. The splendid new steamboat *New England* was lying a few rods off the wharf at Essex, discharging a passenger, when both of her boilers exploded, killing seventeen men and women and scalding many more. "*The New England*'s upper works in the center and after part of the boat were carried away, part of the deck destroyed and several fires started." (Report of jury of inquest.)

2. Fred Erving Dayton, *Steamboat Days*, (New York: Frederick A. Stokes, 1925).

3. See Alvin F. Harlow, *Old Towpaths*.

4. See Stewart Holbrook, *The Old Boston Post Road*.

CHAPTER 3

1. *National Archives*. See "Corporate Records."

2. See Christopher Roberts, *The Middlesex Canal*.

3. With Long Island steamboats arriving from or departing for New York every hour or two, Providence dominated the commercial life of Rhode Island and a great section of eastern Massachusetts. The Blackstone's prosperity was closely tied to that of Providence; merchandise arriving there could be forwarded up the canal and reach its destination in a few hours.

4. Speaking of Commencement as taking place in early September is not an error. It was then (1829) a one-day occasion and not the social affair it has become, lasting a week.

CHAPTER 4

1. Bordentown, New Jersey, was where it eventually reached the Delaware River.

2. Baldwin had become an accomplished surveyor and had taught himself the rudiments of canal engineering. There were wealthy men in Massachusetts

who nurtured the dream of building a canal from Boston to the Connecticut River and on westward to the Hudson. They had engaged him to run a line for such a project. That he used excellent judgment was shown when the Boston and Albany Railroad followed his original survey years later.

3. In either case it entailed a journey of five to six miles. Harlow speaks of gaily decorated oxen and the throngs of country people who followed the wagon to the canal and cheered the launching.

4. The canal needed toll collectors and husky men to handle the heavy balance-beam lock gates. It found them locally. To be able to say he was "working on the canal" soon began to give a young man a bit of social status.

5. The weeks lost in clearing away the debris by hand and wagon doubled the cost of construction. Often clearing the Deep Cut after one fall, a second or even a third occurred.

6. In the *National Archives* there is a lengthy account of how the flotilla of steamboats arrived at Perryville and transported troops down to Annapolis.

CHAPTER 5

1. The magnificent quadruple set of parallel locks permitting the passage of shipping from Lake Superior to Lake Huron, and vice versa, at Sault Ste. Marie, Michigan, handles a freshwater tonnage second to none in the world. It is an unforgettable sight to observe from your hotel window great steamships as they appear to be gliding across town at street level.

2. See Christopher Roberts, *The Middlesex Canal*.

3. Much has been written about the Irish "bogtrotters" who built the Erie Canal—and deservedly so. But the conditions they toiled under—muck up to their waists, hordes of mosquitoes, malaria, poor food, the ground or a slab of wood for a bed—were no worse than what confronted the Negro builders of our southern canals, but no one has mentioned the hardships of these slaves.

4. Harriet Beecher Stowe used the Dismal Swamp for the setting of her anti-slavery novel *Dred*, published in 1856. She mentions Lake Drummond, famous for its amber-colored water, which was prized by ship captains because it remained fresh long after being put in casks.

5. Three times governor of South Carolina; U.S. senator and minister to Spain during the years when Louisiana was relinquished to France and then sold to the United States.

CHAPTER 6

1. Most sources agree that the man who accompanied Washington on this journey was a Negro slave.

2. See Alvin F. Harlow, *Old Towpaths.*

3. *Letters of Abigail Adams*, a lively, piquant commentary on early Washington, perhaps the best we have (American Biographical Series, 2 volumes. New York: Garrett Press, 1969).

4. Construction of the National Road began in 1811 and was completed to Wheeling in 1818. According to the system of John Loudon McAdam, the Scottish engineer whose English coaching roads were famous, the surface was composed of a mixture of sand, gravel and tar which was rolled until bound together. From this came our word *macadam.*

5. A second subscription of $300,000 was made in 1837. Work on the extension began in 1837 and water was turned in on July 4, 1843.

6. During the Civil War few railroads suffered as heavily as the Baltimore and Ohio. In his *A Short History of American Railways*, Slason Thompson, says: "The Confederate forces in May 1861, took possession of more than one hundred miles of the main line, mostly between the Point of Rocks and Cumberland; and by occasional raids caused great destruction on the road between Cumberland and Wheeling . . . locomotives, cars and machinery were carried off and 'transported by animal power' over turnpikes to southern railways."

CHAPTER 7

1. The long, narrow bateaux or flats usually traveled in groups of three or four, an experienced riverman, regarded as a senior captain, directing traffic. They soon became acquainted with everyone on the river and greeted passing boats going in the opposite direction with good-natured taunts. Favorite spots at which to pull up for the night soon were established and it was not unusual for twenty-five men or more to cook their supper together: corn pone, chickens, eggs and bacon.

2. For half a century the turnpike over the mountains was used extensively by emigrants bound for Tennessee and Kentucky. Until West Virginia seized control in 1863, the company had found it profitable. From Gate City it followed the Holston River southwest to Knoxville, Tennessee.

3. William Henry Harrison, the hero of the Battle of Tippecanoe, died of pneumonia on April 4, 1841, thirty days after his inauguration as President of the United States.

4. One of Lexington's great attractions was the limestone arch, 215 feet high and 150 feet wide, a few miles south of town, known worldwide today as Natural Bridge. Thomas Jefferson, says history, bought it for twenty shillings in 1775 and built a cabin there to accommodate visitors.

5. Accompanied by two of his staff Jackson was inspecting his lines at night at Chancellorsville when he was shot by one of his own pickets in the mis-

taken belief that the general and his aides were Union vedettes. Jackson's right arm had to be amputated at the shoulder. In great agony, he suffered for a week and died on May 10.

CHAPTER 8

1.   Horatio Allen was only twenty-seven when he was sent to England. He was to have a distinguished career in engineering. He built the famous reservoir at Fifth Avenue and Forty-second Street in New York where the Public Library now stands. Another of his achievements was the construction of High Bridge which carried the Croton Aqueduct across the Harlem River to New York City. Later he worked with Roebling in building the Brooklyn Bridge. He became chief engineer of the Erie Railroad and in 1846 was elected its president.

2.   When Jervis left the Delaware and Hudson Canal, he took command of the infant Mohawk and Hudson Railroad. The English-made locomotives then in use in this country were tearing up the M. and H.'s light rails. Jervis remedied the situation by "removing the rigid front axle, with its single pair of wheels, and substituted a front truck with two axles and four wheels." The innovation proved to be one of the greatest advancements in the history of the American-type locomotive.

3.   See Edwin LeRoy, *The Delaware and Hudson Canal: A History*.

4.   When steamboats became numerous on the Hudson River, many found Rondout Creek an excellent winter haven, where they were in no danger of being battered by ice when the annual spring breakup occurred. But a score of boats tied up rail to rail were helpless when one caught fire. Many burned to the water's edge.

5.   Horatio Allen, *Railroad Era* (New York: Railroad Gazette, 1884).

6.   At Lake George, then an exclusive summer resort for the very rich, it built a handsome, two-story depot at the edge of the lake. It hasn't heard a train whistle in many years. But the ornate depot still stands, unused and growing shabby with the passing years.

CHAPTER 9

1.   Requesting that forty thousand bushels of coal be produced in twenty years was asking for very little. But it would be sufficient to determine if anthracite was to find a market.

2.   It was and still is the most successful form of pool navigation ever devised. After being emptied, the "bear-trap" dam could be refilled quickly. It enabled boats to pass without delay.

3.   In Easton *Argus*, April, 1862.

CHAPTER 10

1.   These figures are given by Slason Thompson in his *A Short History of American Railways*. Harlow puts the total mileage at three thousand miles.

2.   Among Horatio Allen's credits should be listed the founding of that bulwark of affluence, the Union League Club of New York City.

3.   Young George McCullock, while fishing on Lake Hopatcong, is said to have conceived the idea of a canal connecting the Delaware River with Newark Bay and to have discussed his idea at a meeting at Drake's Tavern, in Morristown, in August, 1822.

4.   The directors of the Morris Canal and Banking Company acquired a $5,000,000 internal improvements bond issue by the state of Michigan at an undisclosed discount, and proceeded to sell the bonds without bothering to pay for them. This was the company's severest scandal.

5.   The first anthracite-burning furnace in the United States was built at Stanhope.

CHAPTER 11

1.   See Stewart Holbrook, *The Story of American Railroads*.

2.   According to Henry V. Poor in *Manual of Railroads and Canals, 1860*: "The Company practices a studious concealment of its affairs. . . . The Company is the paramount authority in the State, dictating legislature [*sic*] upon all subjects."

3.   See Slason Thompson, *A Short History of American Railways*.

4.   It was Alexander Cassatt's vision and perseverance that put the Pennsylvania into New York City. Inscribed on his statue in the great terminal are the words: "Foresight, courage, and ability."

5.   Quoted in *Adventures of America: 1857–1900*, John A. Kouwenhoven, editor (New York: Harper & Bros., 1938).

CHAPTER 12

1.   Credit for building the first ark seen on the upper Susquehanna belongs to Williamson. It was built on the banks of the little Cohocton River, near Bath, New York, and launched in 1800. Although many foundered before reaching their destination, 142 were reported as having reached tidewater at Port Deposit in the 1822 season.

2.   On the upper Conemaugh, east of Johnstown, a large reservoir was built to impound water for the canal. After that division had been abandoned,

hotels and summer villas of wealthy Pittsburghers were built there. It was this dam that gave way on May 31, 1889, the resulting flood which swept down the valley taking the lives of at least twenty-two hundred men, women and children, referred to ever after as the Great Johnstown Flood.

3. Steam locomotives replaced horses on the Portage Railroad several years after it first became operative. In 1850 new surveys were made and work done to eliminate the planes and make the passage over the Allegheny spine by rail.

4. See Charles Dickens, *American Notes*, Vol. II.

CHAPTER 13

1. It boggles the mind to try to estimate how many million tons of coal have been boated down the Susquehanna from its north branch and west fork. Great as that figure must be it is probably exceeded by the splintered, tortured tons of anthracite that have ridden downriver on the current. For half a century and more dredges have been scooping it up from the bottom, screening, washing and selling it to the steel mills. Harrisburgers speak fondly of what they have termed the "Susquehanna Navy."

2. The Pennsylvania Canal Commission, in an effort to slow down the drainage of business away from the state canals and its Philadelphia and Columbia Railroad, unlawfully entered into secret agreements with large western shippers giving them cash rebates in return for forwarding their freight exclusively over state facilities. This, in reverse, was the ugly system employed by Standard Oil many years later to undercut its competitors. In the case of the Canal Commission, the skullduggery was quickly detected and the practice stopped.

3. Jubal Anderson Early, a West Pointer, resigned his commission in 1838 and began the study and practice of law at Rocky Mount, Virginia. He opposed secession but was loyal to his state. In 1864, he led the great Confederate raid down the Shenandoah Valley and suffered a succession of defeats. After three years of self-imposed exile in Mexico and Canada, he returned to Lynchburg and resumed practicing law.

CHAPTER 14

1. Canada was the Lake Champlain region's most important market. It was easily reached by sailing up Lake Champlain and down the Richelieu to the St. Lawrence. What the northeastern counties needed was a connection with the Hudson.

2. The Durham, which originated in England, first appeared in the United States on the Delaware River. Although it rested higher in the water than the American canal boat, which soon superceded it, it had considerable sheer from its rounded bow to stern, which aided its navigation.

3. See Alvin F. Harlow, *Old Towpaths.*

4. It is ironic that New York City, which became the leading metropolis of the United States in the years immediately following the completion of the Erie and owed so much of its booming prosperity to the canal, should have opposed its building at every turn.

5. De Witt Clinton served for two triennial terms as governor of New York (1817-1823). He was again elected to that office in 1825 and served until his death in 1828.

CHAPTER 15

1. The new five-man commission, composed of Clinton, Stephen Van Rensselaer, Samuel Young, Joseph Ellicott and Myron Holley, met in New York City in December. Clinton was elected president, Young made secretary and Holley named treasurer.

2. The Long Level ran from Lock 53 at Frankfort to Lock 54, three-quarters of a mile east of Syracuse in the old village of Salina. It was said to be the longest canal level in the world.

3. Quoted by Noble E. Whitford, foremost Erie Canal authority, in his *History of the Canal System of the State of New York*.

4. See Alvin F. Harlow, *Old Towpaths*, p. 53.

5. Black Rock long ago was incorporated into the city of Buffalo.

CHAPTER 16

1. The type of canal boat encountered in Walter Edmonds' engaging canal fiction. See his *Rome Haul, Mostly Canallers, Erie Water*, etc.

2. See Lionel D. Wyld, *Low Bridge: Folklore of the Erie Canal*.

3. See Noble E. Whitford, *History of the Canal System of the State of New York*.

4. The city of Albany built a dock for canal boats, said to have been "nearly a mile long."

5. William Leete Stone, *Journal of a Tour from New York to Niagara* (Buffalo: Buffalo Historical Society, 1910).

CHAPTER 17

1. Formerly, it had cost $3 to ship a barrel of flour by wagon freight from Rochester to Albany. Via the canal it could be delivered for seventy-five cents. Such a slashing of shipping costs produced a market for all products that had not existed previously. Even baled hay could be landed in Albany at a saleable price.

2. Roberts had taken the precaution of testing the "Combines" before Clinton and his party arrived, but he failed to record the identity of the boat that served as the guinea pig for his great engineering feat.

3. See Stewart Holbrook, *The Erie Canal.*

4. Sam Mitchell and his vials of water from the great rivers of the world has been accepted as factual by most historians. If true, he must have been the better part of a year gathering his samples, world communications being what they were, which the present writer doubts.

5. Quoted by Lionel D. Wyld, *Low Bridge: Folklore of the Erie Canal.*

CHAPTER 18

1. See Joseph Austin Durrenberger, *A Study of the Toll Road Movement in the Middle Atlantic States and Maryland.*

2. Playing the title role of Tess in the motion picture version of the novel made Mary Pickford, a hitherto unknown young actress, famous.

3. The official population of Ohio as given in the 1830 census was 937,903.

4. See Noble E. Whitford, *History of the Canal System of the State of New York.*

5. A serious break was mended by driving double rows of stakes into the earth, interlacing them with planks, filling the break with rocks and twists of hay, and then plastering the whole with a heavy, waterproof covering of clay.

6. At the hydraulic locks a boat and its cargo were weighed as one. By subtracting the registered weight of the boat from the gross weight, toll charges on the cargo were quickly arrived at.

7. See Charles F. Carter, *When Railroads Were New.*

CHAPTER 19

1. Massillon prospered and left Canton far behind. In an attempt to recover lost ground, Canton wasted $10,000 building a branch canal to connect with the Sandy and Beaver.

2. The Toledo and Hocking Valley Railroad became one of the most prosperous coal carriers in the state and over leased lines put its trains through to Lake Erie.

3. Prior to 1862 Cincinnati was the principal center in the United States for the slaughtering of hogs and the packing of pork. It began in 1835 with the coming of thousands of German immigrants who settled in the district northeast of the Miami and Erie Canal, which quickly became known as "over the Rhine."

4. The improvements that had been made on the Muskingum were sold to the federal government in 1887 as part of its flood control program.

5. The Grand Reservoir was breeched several times. The perpetrators were known, but a jury could not be empaneled to try them.

CHAPTER 20

1. To avoid confusion for those readers unacquainted with Indiana, it should be pointed out that the Whitewater and White River are different streams.

2. Indianapolis was a one-street village of only six hundred inhabitants when it became the capital of Indiana. The ground on which the state buildings stood was a grant from the federal government.

3. See Madeline Sadler Waggoner, *The Long Haul West*, pp. 244-45.

CHAPTER 21

1. General Lewis Cass was the first governor of Michigan Territory, which at the time included part of what later became Indiana. Many honors were bestowed on him. He was in turn minister to France, secretary of war, U.S. senator and candidate for President of the United States.

2. The Chicago, New Albany and Louisville was to become that beloved Hoosier Railroad, known to all as the Monon. One of its famous passengers was General Lew Wallace, traveling back and forth between his home in Crawfordsville and Indianapolis. Monon officials like to think that Wallace penned part of his novel *Ben Hur* on its trains.

CHAPTER 22

1. The first *J. M. White* was built at Elizabethtown, Pennsylvania, and made Billy King, her designer, famous. She was 250 feet long, with a beam of 37 feet and draft of 8 feet. After she had been destroyed by fire, a second and then a third *White* appeared. They were fine boats but rivermen never regarded them with the affection they had felt for the original.

2. Captain Kountz was commissioned a commodore at the outbreak of the Civil War by the War Department and placed in charge of river transportation. In that capacity he purchased the dozens of steamboats that were converted into gunboats and troop carriers.

CHAPTER 23

1. Alfred T. Andreas, *History of Chicago from the Earliest Period to the Present Time,* 3 vols. (Chicago: A. T. Andreas, 1884-86). See Vol. I.

2. See Elias Colbert, *Chicago and the Great Conflagration.*

3. Against its wishes, William Butler Ogden gave Chicago its first railroad, the little Galena and Chicago Union. In November, 1848, he brought into the city

from Des Plaines, ten miles away, the first carload of wheat to reach it by rail. A week later ten more carloads of wheat arrived. It created a sensation, embarrassing those politicians who had denied Ogden the right to establish a freight depot on the waterfront on the grounds that farmers preferred to bring their grain to town by wagon. From that humble beginning came the great Chicago and Northwestern Railroad System.

4. Francis Sherman was twice elected mayor of Chicago.

5. Passenger fare from Chicago to La Salle was set at $4, including meals, which amounted to about two cents a mile. This was in line with fares on most eastern canals. When the *John Kinzie* made its fast run from La Salle to Chicago in twenty-two hours, the competition matched it in their advertisements if not on the canal.

CHAPTER 24

1. Voters had been assured that the lottery would provide sufficient funds to defray the cost of public education in New York State. That goal has not yet been reached, due in part to other forms of legalized gambling and the wide sale of Irish Sweepstakes tickets.

2. General Beauregard died in 1893, in New Orleans, while serving as commissioner of public works.

CHAPTER 25

1. See Charles Dickens, *American Notes*, Vol. II.

2. Locks were drained with the close of navigation for the year, an event that was eagerly awaited by village boys for it presented them with a day or two of interesting scavenging which was often rewarded by some interesting finds.

3. See Lionel Wyld, *Low Bridge: Folklore of the Erie Canal*, Edward Noyes Westcott, *David Harum,* Irving Bacheller, *Eben Holden.*

CHAPTER 26

1. Nationally we knew very little about the "North Country" until Irving Bacheller published his novel *Eben Holden* in 1900.

2. The number of knockdowns registered in a fight was not necessarily a measure of its savagery, for the knockdown was used by the ring-wise boxer as a stratagem to clear his wits or recover his wind. Simply by slipping to one knee he was awarded ten seconds in which to decide whether to quit or carry on.

3. Demonstrating the loyalty of Erie men to their home country, Pullman returned to Palmyra, New York, to build his first cars after organizing the Pullman Palace Car Company.

4. See Lucius Beebe and Charles Clegg, *U.S. West: The Saga of Wells Fargo.*

A Selected Bibliography

Adams, Samuel Hopkins. *Canal Town*. New York: Random House, 1944.

————. *Grandfather Stories*. New York: Random House, 1955.

Andrews, Edward D. *The People Called Shakers*. New York: Oxford University Press, 1933.

Andrist, Ralph K. *The Erie Canal*. New York: American Heritage, 1964.

Beebe, Lucius, and Clegg, Charles. *U.S. West: The Saga of Wells Fargo*. New York: E. P. Dutton, 1949.

Bobbe, Dorothie. *De Witt Clinton*. New York: Minton, Balch, 1933.

Carmer, Carl. *The Hudson*. New York: Farrar and Rinehart, 1939.

Carter, Charles F. *When Railroads Were New,* 4th ed. New York: D. Appleton, 1926.

Caruso, John A. *The Great Lakes Frontier*. Indianapolis: Bobbs-Merrill, 1961.

Chalmers, Harry. *How the Irish Built the Erie*. New York: Bookman Assoc., 1965.

Clowes, Ernest S. *Shipways to the Sea*. Baltimore: Williams and Wilkins, 1929.

Colbert, Elias. *Chicago and the Great Conflagration*. New York: Brown, 1872.

Comstock, Howard Payne. *History of Canals in Indiana*. Indianapolis: Indiana Historical Society, 1911.

Dickens, Charles. *American Notes,* Vol. II. London: Chapman and Hall, 1842.

Durrenberger, Joseph Austin. *A Study of the Toll Road Movement in the Middle Atlantic States and Maryland*. Valdosta, Ga., 1931.

Edmonds, Walter. *Rome Haul*. Boston: Little, Brown, 1929.

————. *Mostly Canawlers*. Boston: Little, Brown, 1932.

————. *Erie Water*. Boston: Little, Brown, 1934.

Ellis, David. *A Short History of New York State*. Ithaca, N.Y.: Cornell University Press, 1957.

Federal Writers Project. *The Intracoastal Waterway*. Washington, D.C.: Government Printing Office, 1937.

Fite, Emerson D. *Canal and Railroad from 1861-65*. New Haven, Conn.: Yale Review, 1907.

Goodrich, Carter, ed. *Canals and American Economic Development*. New York: Columbia University Press, 1961.

Hadfield, Charles. *The Canals of Southern England*. New York: Taplinger, 1968.

Hall, Captain Basil. *Travels in North America*. Edinburgh: Lumpkin and Marshall, 1829.

Hanson, Marcys Lee. *The Immigrant in American History*. Cambridge, Mass.: Harvard University Press, 1940.

Harlow, Alvin F. *Old Towpaths*. New York: D. Appleton, 1926.

———. *Old Post Bags*. New York: D. Appleton, 1928.

———. *Old Waybills*. New York: D. Appleton, 1934.

Hepburn, A. Burton. *Artificial Waterways of the World*. New York: Macmillan, 1914.

Hill, Henry Wayland. *Historical Review of Waterways and Canals Constructed in New York State*. Buffalo: Buffalo Historical Society, 1908.

Holbrook, Stewart H. *The Story of American Railroads*. New York: Crown, 1947.

———. *The Yankee Exodus*. New York: Macmillan, 1950.

———. *The Erie Canal*. New York: Random House, 1952.

———. *The Old Boston Post Road*. New York: McGraw-Hill, 1962.

Hosack, David. *Memories of De Witt Clinton*. Boston: L. C. Page, 1941.

Howells, William Dean. *Years of My Youth*. New York: Harper & Bros., 1916.

Hungerford, Edward. *Men and Iron*. New York: Crowell, 1938.

Jennings, W. W. *A History of Economic Progress in the United States*. New York: Crowell, 1926.

Jones, Alexander. *The Chicago Drainage Canal and Its Forebear, the Illinois and Michigan Canal*. Springfield, Ill.: Illinois State Historical Society, 1906.

Kimball, Francis P. *New York: The Canal State*. Albany: Argus Press, 1958.

Kirkwood, James J. *Waterway to the West, James River and Kanawha*. Richmond, Va.: Eastern National Parks and Monuments Association, 1963.

LeRoy, Edwin. *The Delaware and Hudson Canal: A History*. Honesdale, Pa.: Wayne County Historical Society, 1950.

McCullough, Robert, and Leuba, Walter. *The Pennsylvania Main Line Canal*. Martinsburg, Pa.: Morrison's Cove Press, 1960.

Mesuak, J. L. *The English Traveler in America—1785-1835*. New York: Columbia University Press, 1922.

Meyer, B. H. *History of Transportation in the United States before 1860*. Washington, D.C.: Carnegie Institute, 1917.

Morrison, S. E. *The Maritime History of Massachusetts, 1783-1860*. Boston: Houghton Mifflin, 1921.

Niles' Weekly Register. Baltimore: 1830-1837.

Niles, Blair. *James River*. New York: Farrar and Rinehart, 1939.

Payne, Robert. *The Canal Builders*. New York: Macmillan, 1959.

Percher, F. A. *The History of the Santee Canal*. Columbia, S.C.: The South Carolina Public Service Authority, 1950.

Rapp, M. A. *Canal Water and Whisky*. New York: Twayne, 1961.

Report: Committee for Preservation of Historic Sights. (N.Y.). Albany: 1968.

Riegel, Robert E. *America Moves West.* New York: Holt, 1956.

Roberts, Christopher. *The Middlesex Canal.* Boston: L. C. Page, 1938.

Shaw, Ronald E. *Erie Water West.* Frankfort, Ky.: University of Kentucky Press, 1969.

Smelzer, Gerald. *Canal Along the Lower Susquehanna.* York, Pa.: York County Historical Society, 1963.

Thompson, Harold W. *Body, Boots and Britches.* Philadelphia: J. B. Lippincott, 1939.

Thompson, Slason. *A Short History of American Railways.* New York: Appleton, 1925.

Thwaites, Reuben Gold. *Jesuit Relations.* Translated by Edna Hunter. New York: A. and C. Boni, 1925.

Trollope, Frances. *Domestic Manners of the Americans.* New York: Dodd, Mead, 1832.

Veit, Richard F. *The Old Canals of New Jersey.* Little Falls, N.J.: Geographical Press, 1963.

Waggoner, Madeline Saddler. *The Long Haul West: The Great Canal Era, 1817–1850.* New York: Putnam, 1958.

Ward, George Washington. *Early Development of Chesapeake and Ohio Project.* Baltimore: Johns Hopkins University, 1899.

Whitford, Noble E. *History of the Canal System of the State of New York.* Albany: Brandow Printing Co., 1914.

Wyld, Lionel D. *Low Bridge: Folklore of the Erie Canal.* Syracuse, N.Y.: Syracuse University Press, 1962.

Index

Italic page numbers indicate illustrations.